CHINA 2020

CHINA 2020 SERIES

China 2020:
Development Challenges in the New Century

Clear Water, Blue Skies:
China's Environment in the New Century

At China's Table:
Food Security Options

Financing Health Care:
Issues and Options for China

Sharing Rising Incomes:
Disparities in China

Old Age Security:
Pension Reform in China

China Engaged:
Integration with the Global Economy

CHINA 2020

DEVELOPMENT CHALLENGES IN THE NEW CENTURY

THE WORLD BANK
WASHINGTON, D.C.

Manufactured in the United States of America
First printing September 1997

Cover photograph by Dennis Cox/China Stock.

Cover insets (from left to right) by Vince Streano/Aristock, Inc.; Claus Meyer/Black Star; Serge Attal/Gamma Liaison; Dennis Cox/China Stock; Joe Carini/Pacific Stock; Erica Lansner/Black Star.

ISBN: 0-8213-4042-5

Contents

This report uses *Hong Kong* when referring to the Hong Kong Special Administrative Region, People's Republic of China.

Acknowledgments

This report was written by a team led by Vikram Nehru and comprising Aart Kraay and Xiaoqing Yu. Consultants to the team were Athar Hussain (London School of Economics, STICERD) and Barry Naughton (University of California, San Diego). The team was helped by summer interns Jean Imbs and David Ng. The peer reviewers were Dwight Perkins (Harvard University), Gregory Chow (Princeton University), Peter Harrold, Homi Kharas, and Gene Tidrick. The task assistant was Janet Wyse.

The team owes Nicholas Hope (former Director of the China and Mongolia Department at the World Bank) a special debt of gratitude for conceiving the study and for unceasing support. Other World Bank staff to whom the team is grateful for their contributions and comments are Halsey Beemer, Natasha Beschorner, Eduard Bos, Pieter Bottelier, Tilly Chang, Dipak Dasgupta, Yuri Dikhanov, Sandra Erb, Joseph Goldberg, Daniel Gunaratnam, Bert

Hofman, Janet Hohnen, Shaikh Hossain, Gordon Hughes, E.C. Hwa, Todd Johnson, Bert Keidel, Kathie Krumm, Anjali Kumar, Nathalie Lichtenstein, Feng Liu, Kseniya Lvovsky, Tamar Manuelyan-Atinc, Will Martin, Andrew Mason, Hena Mukherjee, Richard Newfarmer, Al Nyberg, Alan Piazza, Klaus Rohland, Helen Saxenian, Lyn Squire, Lee Travers, Jagdish Upadhyay, Michael Walton, and Shahid Yusuf.

The World Bank team benefited from fruitful discussions with many Chinese government officials and academics who were generous with their time and knowledge. The mission is especially grateful to Zhu Xian, Ding Xianjue, Du Jian, Pan Xiaojiang, Wang Zhen, Wang Weixin, Wu Jinkang, Liang Shuchun, Shi Yaobin (Ministry of Finance); Zheng Xinli, Dai Guiying, Cao Yushu, Liu He, Wang Xiduo, Yang Qingwei, Yao Hong, Zhao Shihong, Wang Jianjun, Wu Qiang, Zhao Shijun, Lu Jiaxiang, Yu Peng, Ma Xiaohe, Li Yulin, Hu Chunli, Ning Jizhe, Xu Lin (State Planning Commission); Wang Dayong, Jing Xuecheng, Yang Zaiping, Xie Ping, Yao Keping, Pei Chuanzhi, Zhao Xianfeng, Chen Xin, Yang Huisheng, Zheng Yaodong (People's Bank of China); Wang Zixian, Zhu Zhiping, Liu Yajun, Wang Yi, Chen Xin, Deng Li, Zhang Liyong, Liu Tianmin, Zou Xiaomin (Ministry of Foreign Trade and Economic Cooperation); Wang Huijong, Li Boxi, Li Shantong, Ding Ningning, Zhao Jinping, Ge Yanfeng (Development Research Center); Li Keping, Wang Haijun, Deng Xianhong, Guo Xiangjun, Jiang Xiaoyun, Deng Ran, Chen Yuyu (System Reform Commission); Ming Ruifeng, Wang Yadong, Liu Danhua, Li Jinghu, Lao Yujun, Li Zhengyu, Lu Yulin (Ministry of Labor); Song Yuzhong, Wei Dong, Qin Yongfa, Wang Xu, Liu Dongsheng, Zhou Shuanghu (State Economic and Trade Commission); Song Tingmin (Ministry of Internal Trade), Ma Lin, Zheng Hua, Jin Dongsheng (State Tax Bureau); Feng Juping, Huang Hongbo (State Administration of Foreign Exchange); Liu Beihua (Ministry of Agriculture); Xuan Zengpei (State Science and Technology Commission); Ye Zhen, Zheng Jingping (State Statistical Bureau); Guo Xiaomin, Xiao Xuezhi (National Environmental Protection Agency); Liu Peilong (Ministry of Health); Sun Ling (State Education Commission); Wang Donghong (State Customs Administration); Chi Jianxin (State Development Bank); Wang Loulin, Zhang Zhuoyuan, Yang Shengming, Pei Changhong, Wang Zhenzhong, Wang Tongsan, Li Yang, Lu Zheng, Han Jun, Jiang Xiaojuan, Chen Zhensheng, Zhao Jingxing, Yu Yongding, Zhang Yuyan, Zheng Yisheng, Yu Dechang, Xu Gengsheng, Li Chenggui, Dong Yisheng (Chinese Academy of Social Sciences); Fan Gang (National Economic Research Institute); Hu Angang (Chinese Academy of Sciences); Wang Chuanlun, Tao Xiang (People's University); Yi Gang (Peking University); and Andrew Sheng, Chad Leechor, Jim Wong, Lin Shoukang (Hong Kong Monetary Authority). The team also benefited from discussions with participants at the State Planning Commission–sponsored seminar on "China: Striving Toward the Year 2020" held March 19–21, 1997 in Beihai, China.

The team would like to express its thanks to the Canadian International Development Agency for a generous grant to finance three background papers by Hu Angang (Tsinghua University), Cai Fang (Chinese Academy of Social Sciences), and the State Science and Technology Commission.

Liu Dusheng (World Bank Resident Mission in China) provided coordination and logistical help, while Mei Hong, Fan Wenzhong, and Pan Wenxin (World Bank Department, Ministry of Finance) coordinated arrangements on the government side.

Rupert Pennant Rea edited the report. Bonita Brindley provided editorial guidance. Jennifer Solotaroff helped with production. Meta de Coquereaumont, Paul Holtz, and Glenn McGrath of the American Writing Division of Communications Development Incorporated handled production and layout of the report, and Kim Bieler designed it.

CHINA 2020

China is in the throes of two transitions: from a command economy to a market-based one and from a rural, agricultural society to an urban, industrial one. So far both transitions have been spectacularly successful. China is the fastest-growing economy in the world, with per capita incomes more than quadrupling since 1978. In two generations it has achieved what took other countries centuries. For a country whose population exceeds that of Sub-Saharan Africa and Latin America combined, this has been a most remarkable development.

But every silver lining has a cloud. Swift growth and structural change, while resolving many problems, have created new challenges: employment insecurity, growing inequality, stubborn poverty, mounting environmental pressures, and periods of macroeconomic instability stemming from incomplete reforms. Unmet, these challenges could undermine the sustainability of growth, and China's promise could fade.

This report argues that China can meet these challenges and sustain rapid growth. Although the difficulties ahead should not be underestimated, neither should China's strengths—relative stability, a remarkably high savings rate, a strong record of pragmatic reforms, a disciplined and literate labor force, a supportive Chinese diaspora, and a growing administrative capacity. These strengths have driven China's growth for the last two decades of this century. They could do the same in the first two decades of the next.

But to nurture these strengths and use them effectively, reforms must develop in three related areas. First, the spread of market forces must be encouraged, especially through reforms of state enterprises, the financial system, grain and labor markets, and pricing of natural resources. Second, the government must begin serving markets by building the legal, social, physical, and institutional infrastructure needed for their rapid growth. Finally, integration with the world economy must be deepened by lowering import barriers, increasing the transparency and predictability of the trade regime, and gradually integrating with international financial markets.

chapter one

Understanding the Present

China is in the midst of two historic transitions: from a rural, agricultural society to an urban, industrial one and from a command economy to a market-based one. The first transition would be unremarkable were it not for China's vast size, its past control of urbanization, and the unprecedented speed of its industrialization. Its population easily exceeds the combined total of Latin America and Sub-Saharan Africa, and its industrial growth is nearly an order of magnitude greater. Its transition to a market economy has also been unique, with a combination of experimentation and incremental reforms leading to rapid progress in some areas and slow progress elsewhere.

The interplay and synergy between these two transitions have sparked rapid growth. The Chinese economy expanded more than fourfold in the past fifteen years. Between 1978 and 1995 real GDP per capita grew at the blistering rate of 8 percent a year and lifted 200 million

Chinese out of absolute poverty.[1] Economic reforms, begun in 1978, have advanced China's integration with the world economy, maintained a strong external payments position, essentially privatized farming, liberalized markets for many goods and services, intensified competition in industry, and introduced modern macroeconomic management.

At the same time, the confluence of these two transitions has generated powerful vortices and cross-currents that are potentially destabilizing and always difficult to predict. Transition from command to market alone can be treacherous: witness the period of economic collapse in Eastern Europe and the former Soviet Union. Similarly, transition from a rural, agricultural society to an urban, industrial one also carries many risks. In the rich industrial economies this transition took centuries. In China the process is being telescoped into one or two generations.

So it is hardly surprising that China's rapid growth and structural change, while resolving many problems, have also given rise to new challenges: periods of macroeconomic instability stemming from partially complete reforms; increased employment and income insecurity; mounting environmental pressures, especially in urban areas; rising costs of food self-sufficiency; growing inequality and stubborn poverty levels; and a prickly, occasionally hostile world environment. These are formidable challenges. Unmet, they could undermine the sustainability of growth, and China's promise could fade.

This report argues, however, that China has the capacity to meet these challenges. While the difficulties ahead should not be underestimated, neither should China's strengths—relative stability, a remarkably high savings rate, a strong track record of pragmatic reforms, a supportive Chinese diaspora, and a growing administrative capacity. These strengths have driven China's impressive growth in the last two decades of this century. They will be equally necessary for rapid, sustainable growth in the first two decades of the next. To improve China's chances of success, these strengths will need to be supplemented by three other factors: skilled economic management, a supportive world economy, and domestic social stability.

Swift growth

Measuring China's GDP growth is tricky (box 1.1).[2] Official statistics show that per capita GDP growth averaged 8 percent a year between 1978 and 1995; other estimates suggest growth of 6–7 percent. Still, whatever the precise figure, China's growth has been almost unprecedented: only the Republic of Korea and Taiwan, China, have grown at comparable rates.

China's growth since 1978 is also outstanding in historical perspective (figure 1.1). Performance during the nineteenth century and the first half of the twentieth century could hardly have been more disappointing: the country registered no improvements in living standards between 1820 and 1870 and only modest growth over the next three decades.[3] During the tumultuous first half of this century per capita incomes actually fell as China saw the collapse of a 3,000-year-old imperial tradition, weathered a period of anarchy and warlordism, and survived foreign occupation and civil war. The rest of the world, though, managed respectable growth rates during this period despite the upheaval of two world wars. As a result, China's share in global GDP fell from a high of 30 percent in 1820—the largest economy in the world at the time—to barely 7 percent in 1950.

Since then China has been recovering some of this lost ground. Postwar recovery under communist rule saw a resumption in growth, although the Great Leap Forward and the Cultural Revolution caused considerable economic disruption and waste. By 1978, when China took its first tentative steps toward reforms, it

FIGURE 1.1

China's growth has been outstanding in the second half of this century

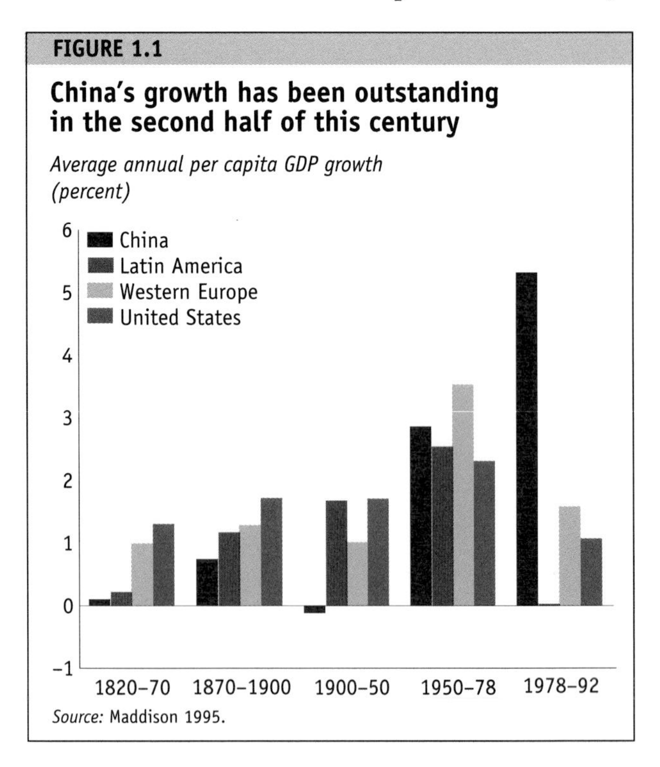

Source: Maddison 1995.

BOX 1.1

How fast is China growing?

According to official statistics, China's per capita GDP grew 8 percent a year (in real terms) between 1978 and 1995. This figure may overstate China's performance, however, because the official consumption and investment deflators used to convert nominal into real GDP have increased much more slowly than alternative measures, such as the consumer price index or price indexes for capital goods. If these alternative indexes are better measures of price increases, they can be used to deflate nominal GDP to arrive at a more accurate measure of real improvements in living standards.

Using reasonable alternative measures of price increases to deflate nominal GDP indicates that per capita growth may be 1.2 percentage points lower than is indicated by official statistics (see table). Most of the discrepancy between the two estimates arises in the second half of the reform period (1986–95).

How significant are these adjustments? On the one hand, mismeasuring GDP growth by 1 or more percentage points over long periods will result in large errors in per capita income levels. But even after data deficiencies are taken into account, China's growth performance remains stellar (see figure). By any measure, China remains comfortably among the world's ten fastest-growing economies.

Deflating China's growth

(annual per capita GDP growth, percent)

	1978–95	1978–86	1986–95
Using official deflators	8.0	7.8	7.9
Using alternative deflators	6.8	7.4	6.6

Note: See note 2 at end of chapter.
Source: China Statistical Yearbook, various issues; World Bank staff estimates.

China's record growth, 1978–94

Annual per capita real GDP growth (percent)

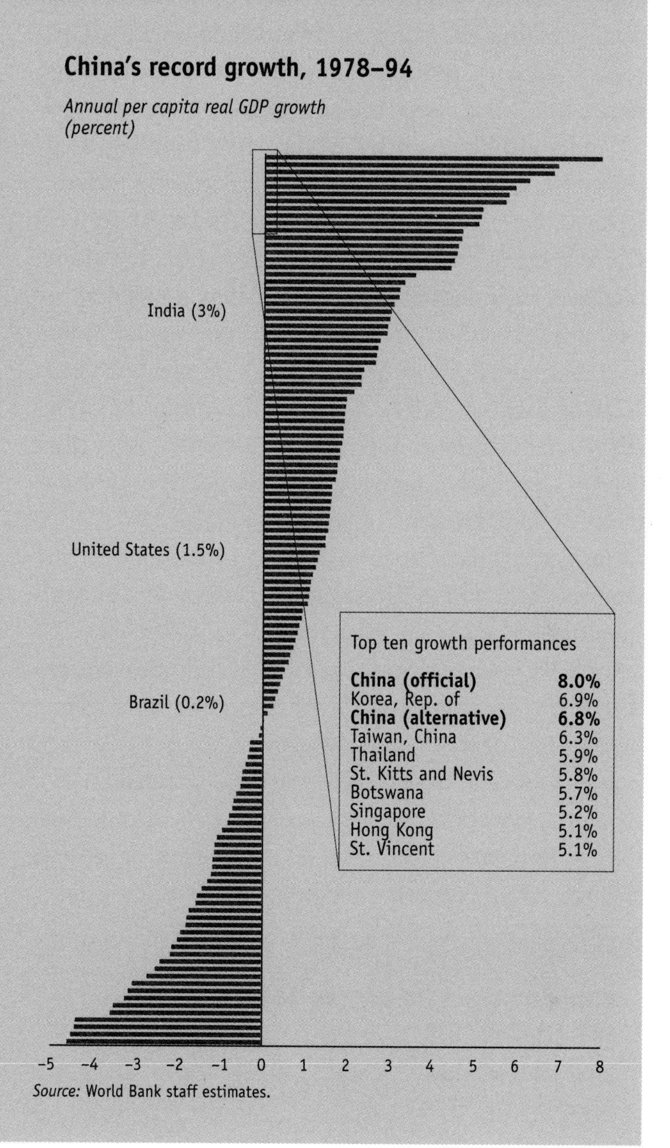

Source: World Bank staff estimates.

remained a desperately poor, rural, and agricultural economy. Sixty percent of China's 1 billion citizens survived on less than $1 a day, the international poverty standard.

The rest, as they say, is history. The introduction of agricultural reforms was the first sprinkling that rejuvenated the parched economy, launching a growth process that has transformed the face of China. The torrid growth rate set over the next seventeen years helped the Chinese double their per capita income every ten years, faster than almost any country in recent history (figure 1.2).[4]

Three features of China's rapid growth are especially noteworthy. The first is its regional dimension. Despite concerns about regional disparities, the benefits of growth have been widely shared among China's country-size provincial economies. Although the coastal provinces grew faster than average at 9.7 percent a year, the other provinces also fared well. Indeed, if China's thirty provinces were counted as individual economies, the twenty fastest-growing economies in the world between 1978 and 1995 would have been Chinese.

The second notable feature is the sharp cyclical pattern of economic growth. Although the ups and downs of China's postreform business cycles pale in comparison with the wild swings in GDP growth seen during the Great Leap Forward and the Cultural Revolution, they nevertheless have been wide enough to concern policymakers (figure 1.3). These growth cycles have been accompanied by similar fluctuations in the rate of inflation, revealing fault lines in macroeconomic management stemming from partially completed reforms in the fiscal, enterprise, and banking systems.

The third noteworthy feature of China's growth since 1978 is its reliance on productivity growth. Relative to other rapidly growing Asian economies, China's growth has been less dependent on volume increases in inputs of capital and labor. Consider, for example, growth in the stock of physical capital (table 1.1). In most countries growth in capital inputs exceeds GDP growth, often by a substantial margin.[5] In China the reverse occurred, suggesting that factors other than capital accumulation have been important determinants of GDP growth.

In fact, a conventional technique to account for the sources of growth in China reveals that growth in capital inputs explains just 37 percent of growth (annex 1). Another 17 percent can be attributed to improvements in the quantity and quality of the labor force. Thus nearly half of China's GDP growth—4.3 percentage points a year—is due to other factors. Data deficiencies along the lines of those discussed in box 1.1 can reduce the absolute size of this residual component of growth slightly, as can differing methodological assumptions. But even after these adjustments, the fact remains that China's growth has come from much more than the mere accumulation of factors of production.

China's remarkably rapid growth since 1978 has been driven by four factors:

- A high savings rate, which has supported vigorous rates of investment and capital accumulation.
- Structural change, which has been both a cause and an effect of growth.
- Pragmatic reforms, which were well suited to China's unusual circumstances and enjoyed broad support.
- Economic conditions in 1978, which were especially receptive to reform; China's economy could be described as a dry prairie, parched by years of planning, awaiting the first sprinklings of market reform.

Each of these factors is considered in turn.

High savings

The most striking feature of China's remarkable performance since 1978 is its savings rate. Savings, as much as growth, is China's real economic miracle. According to official statistics, China's savings rate averaged 37 percent of GDP between 1978 and 1995, although more conservative estimates place it at 33–34 percent.[6] Even so, this rate is among the highest in the world.

No less important, the savings rate was remarkably stable, even as reforms and structural change were reshaping the economy. Contrast this with the collapse of savings in the transition economies of Eastern Europe and the former Soviet Union (figure 1.4). In fact, the sta-

FIGURE 1.2

China doubled its per capita income at a remarkable rate

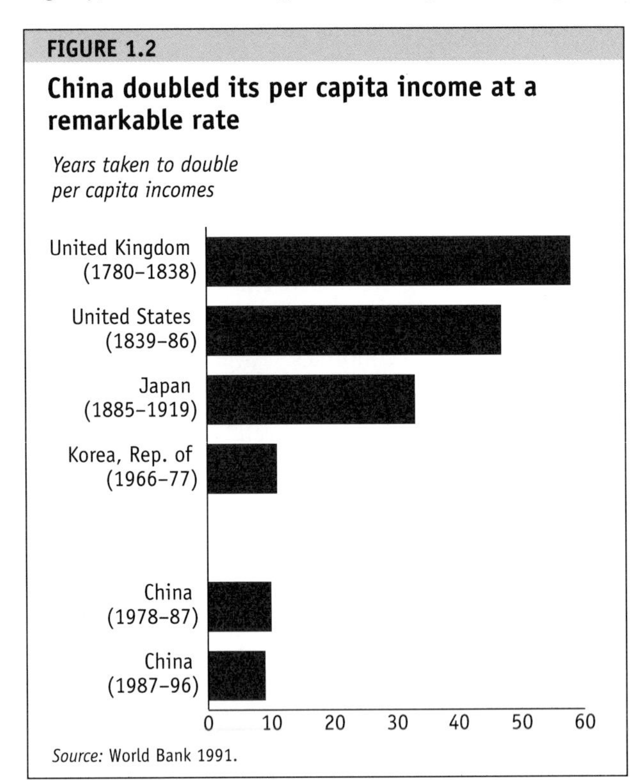

Source: World Bank 1991.

FIGURE 1.3

The ups and downs of China's growth, 1953–93

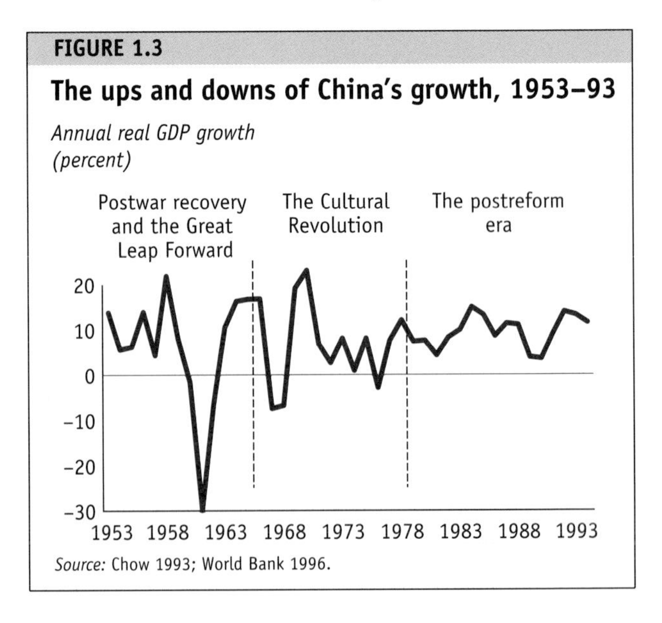

Source: Chow 1993; World Bank 1996.

TABLE 1.1
Accounting for China's growth
(percent)

Country	Period	Average annual growth: GDP	Physical capital	Human capital	Labor force	GDP growth explained by sectoral reallocation	Unexplained share of growth
China	1978–95	9.4	8.8	2.7	2.4	1.5	29
Comparators							
United States	1820–1913	4.1	5.5	1.6	2.8	n.a.	14
United States	1950–92	3.2	3.2	1.1	1.6	0.03	35
Japan	1960–93	5.5	8.7	0.3	1.0	0.26	30
Korea, Rep. of	1960–93	8.6	12.5	3.5	2.4	0.34	21

Note: Human capital is defined as stock of years of education per worker. Sectoral reallocation captures the shift of resources from low- to high-productivity sectors. In China it also includes reallocation from state to nonstate industry. See annex 1 for a complete description of methodology, data sources, and adjustments made to official Chinese data.
Source: Annex 1 of this report.

bility of the high savings rate was one of the primary successes of the reform path chosen by the Chinese.

Although China's savings performance sets it apart from other transition economies, it fits more comfortably in the mold of other dynamic Asian economies. All had rapidly rising savings rates in the periods immediately following takeoff, contributing to virtuous circles of high growth and high savings. China appears to be following this pattern, although its extremely high savings rate at a very low income level makes it an exception even within this select group (figure 1.5).

What accounts for China's high savings rates? Under central planning high savings rates were engineered through the plan. Agricultural and raw material prices were kept artificially low and final goods prices artificially high. This concentrated profits in state enterprises, which invested them under the direction of the central and local planning apparatus. Between 1965 and 1978 savings remained at 33 percent of GDP, and investment strongly favored heavy industry. Household savings were of little importance, averaging just 1 percent of GDP.

Since reforms began in 1978, however, the roles of enterprises and households have been reversed. The household savings rate exploded from about 1 percent before reforms to 21 percent since then (figure 1.6). Households now contribute half of total savings.

FIGURE 1.4
China's other miracle: Savings during transition

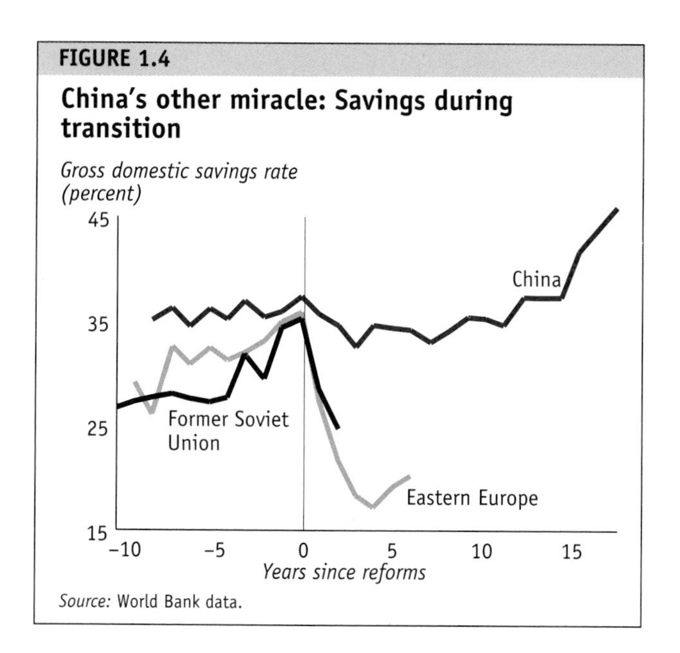

Source: World Bank data.

FIGURE 1.5
China's savings rate: Apart, even from a select crowd

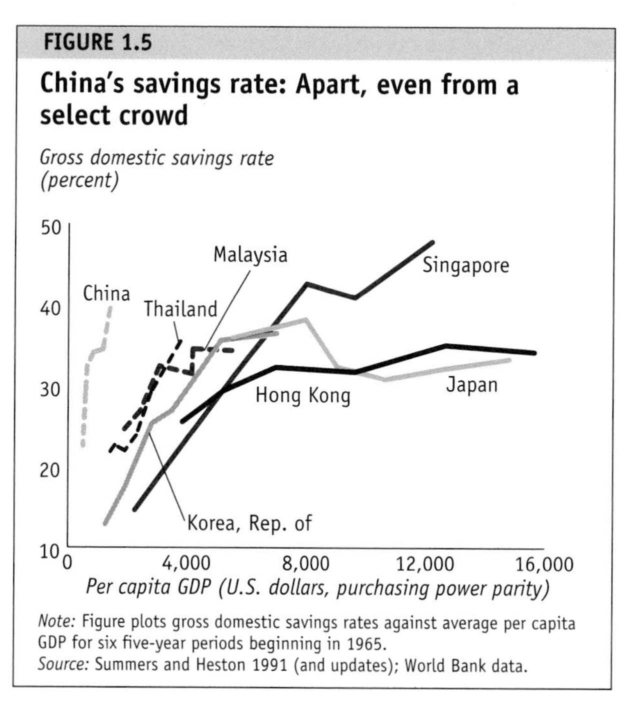

Note: Figure plots gross domestic savings rates against average per capita GDP for six five-year periods beginning in 1965.
Source: Summers and Heston 1991 (and updates); World Bank data.

Why are Chinese households so thrifty? No single explanation seems adequate. Certainly, for many Chinese, rising incomes have brought an end to subsistence consumption. Rising incomes have also been accompanied by rising aspirations, as households set their sights on buying their own house or apartment, acquiring previously unavailable consumer durables, and giving their children a better education. Since households could not borrow much to finance these purchases, they saved more instead.

Another factor is China's changing demographics.[7] Life expectancy today is ten years longer than it was in 1970, and it continues to rise. A typical Chinese baby can now look forward to living for more than seventy years. Because city dwellers retire at 55, they have strong incentives to save for their retirement, especially when they know that pension benefits may be inadequate. Furthermore, declining fertility has reduced the traditional form of support in old age—children.

Finally, institutional factors have also helped boost household savings. Of these, two stand out: the implicit public guarantee of deposits in the banking system and the proliferation of new ways to invest in China's fledgling securities markets.

Structural change

With swift growth came rapid structural change. In this respect China followed the path of many other countries.[8] Where China differed was in the pace of change: it compressed into a few years a process that has normally taken several decades. Consider employment patterns. In the eighteen years since 1978, agriculture's share of the workforce dropped from 71 percent to about 50 percent. It took the United States fifty years and Japan sixty years to achieve a similar structural shift (figure 1.7).

"Push" and "pull" factors accelerated the flow of labor out of agriculture. Low incomes from farming and widespread poverty in rural areas encouraged farmers and their families to leave.[9] At the same time, the demand for labor increased sharply in industry and services—especially among collectively owned enterprises—that had achieved rapid productivity growth.

The shift out of agriculture did not mean massive migration to cities. Cities' share of the population increased from 18 percent in 1978 to just 29 percent in 1995 (box 1.2). Compared with the period before reform, however, urbanization was rapid. Between 1957 and 1978, when official policy discouraged urbanization, the share of the population living in urban areas climbed only 3 percentage points, from 15 to 18 percent.

The drift away from agriculture also facilitated a transformation in patterns of ownership, especially in industry. The first wave of industrialization took place after the establishment of the People's Republic and was concentrated in state enterprises. Since 1978, however, these enterprises have grown relatively slowly, so their share in aggregate output and employment has been falling swiftly. The second wave of industrial growth was in collectively owned and township and village enterprises. In the past few years a third wave of industrialization has been building: mostly privately and

FIGURE 1.6

China's thrifty households

Household savings as a share of income (percent; five-year moving average)

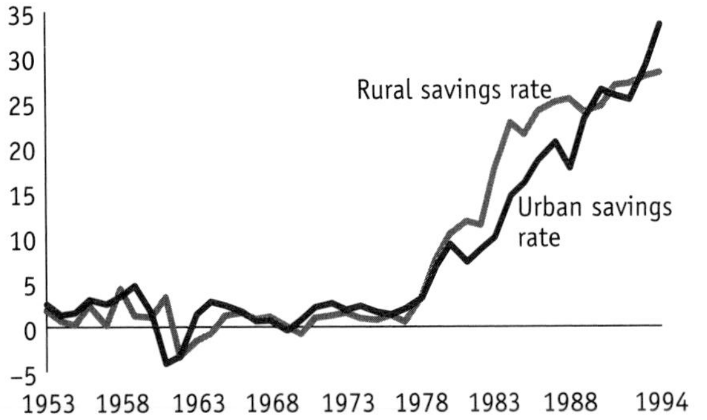

Rural savings rate

Urban savings rate

35 30 25 20 15 10 5 0 -5

1953 1958 1963 1968 1973 1978 1983 1988 1994

Source: China Statistical Yearbook, various issues.

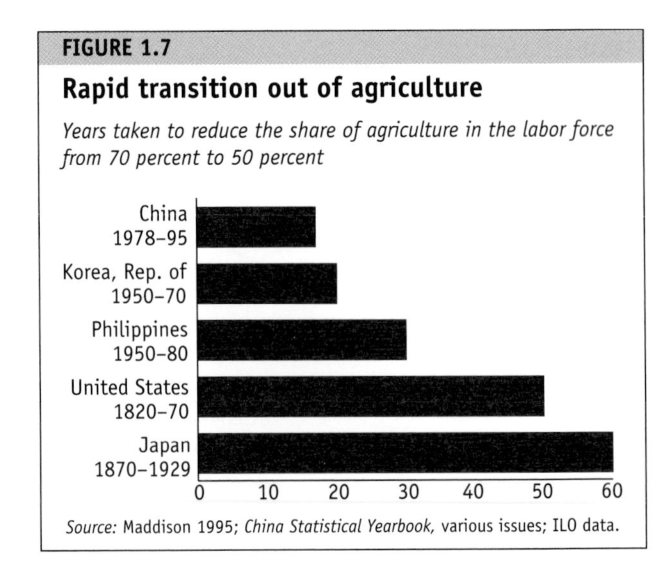

FIGURE 1.7

Rapid transition out of agriculture

Years taken to reduce the share of agriculture in the labor force from 70 percent to 50 percent

Source: Maddison 1995; *China Statistical Yearbook,* various issues; ILO data.

BOX 1.2

How urbanized is China?

Despite the recent expansion of cities and towns, China's urban population—29 percent of the total, according to official statistics—is still below the 35 percent norm for countries with similar per capita incomes (calculated on a purchasing power parity basis). But China's urban population statistics are based on administratively defined boundaries, some of which include low population-density areas engaged in farming. At the same time, large tracts of coastal China's so-called rural areas are developing a new, dispersed pattern of urbanization. In three regions—the Pearl River delta, the lower Yangtze, and the Beijing-Tianjin-Tanshan metropolitan areas—industrial and commercial activities are transforming the landscape, forming urban and periurban corridors linking core cities. If regions with high population densities are considered urban areas, China's level of urbanization could be as high as 50 percent, significantly above that of the typical low-income country.

Source: Cai 1996.

individually owned enterprises, supplemented by joint ventures and foreign-funded enterprises.[10]

If anything, the size of the private sector is larger than official statistics suggest. Many private firms have reason to underreport their production and employment. Moreover, many collectively owned enterprises function as private enterprises in virtually every respect but find it convenient to operate under the banner of townships or villages—a practice known as "wearing the red cap." Doing so may help the enterprises obtain access to credit or licenses and avoid social or political stigmas that may still cling to private activity in China.

Structural change has given an extra boost to China's growth over the past eighteen years. Since a large portion of the agricultural labor force was underemployed, productivity leaped as workers moved from low-productivity agriculture to more productive employment in industry and services. Between 1978 and 1995 this process contributed about 1 percentage point a year to GDP growth. The changing pattern of ownership also contributed about 0.5 percentage point a year to growth, as employment shifted to the collective and private sectors, where productivity was higher (annex 1).

The process of structural change varied across provinces, exacerbating interprovincial and especially rural-urban income differences (figure 1.8).[11] In provinces that industrialized faster, higher wages in

FIGURE 1.8

Provincial growth, structural change, and inequality

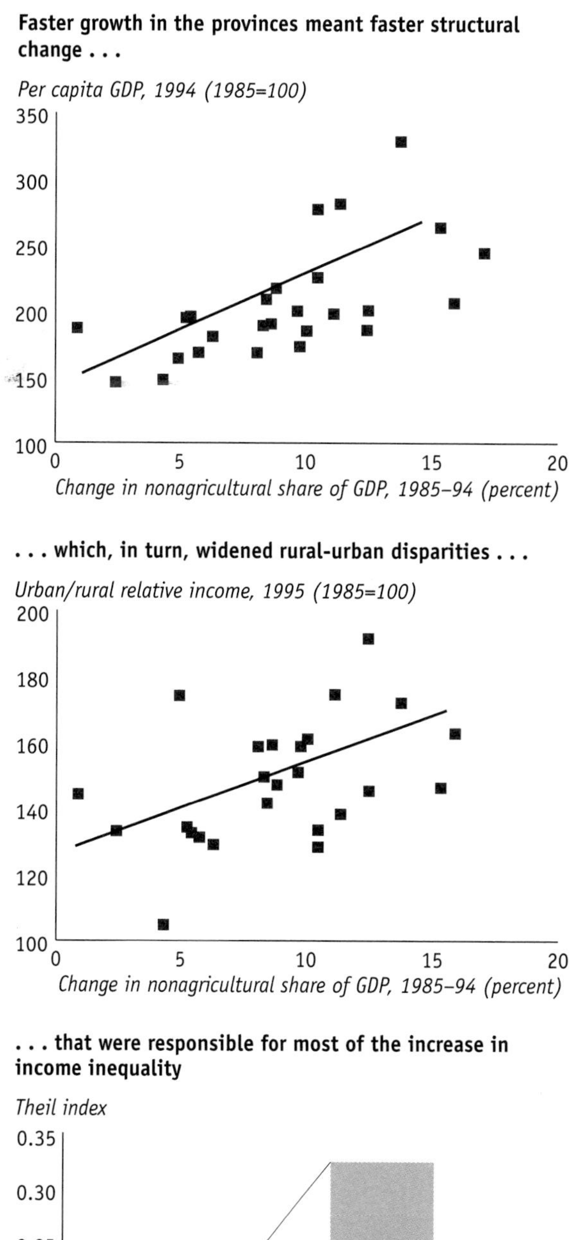

Note: See note 11 at end of chapter.
Source: World Bank staff estimates.

urban and periurban areas opened a growing gap between urban and rural workers. Restrictions on rural-urban migration helped maintain this gap, which was responsible for much of the sharp increase in income inequality over the past decade. Today urban incomes are as much as four times rural incomes, once subsidies enjoyed by urban residents are taken into account. Such rural-urban gaps are very high by international standards. Evidence from thirty-six countries shows that the ratio of urban to rural incomes tends to be below 1.5 and rarely exceeds 2.0.[12] In China the gulf between rural and urban residents explains 60 percent of overall income inequality. In contrast, regional income disparities have contributed relatively little to inequality (figure 1.9).

Still, China is more egalitarian than most countries in Africa, Latin America, and even East Asia. But China's steep rise in inequality is exceptional in international perspective (figure 1.10). In the early 1980s China's income inequality was well below average in a sample of forty countries for which data are available. By the 1990s it was above average.

Pragmatic and incremental reforms

China's economic reforms in 1978 were triggered by neither economic crisis nor ideological epiphany. The country had endured much hardship over the previous two decades, with the start of the Great Leap Forward and through the Cultural Revolution. Against this tumultuous backdrop the years leading up to 1978 were relatively tranquil.

The Chinese leadership was, therefore, eager to see improvements in living standards but had no appetite for dramatic changes in policy. Growth was important, but not at the expense of stability. The concern with growth initially focused on restoring incentives for agricultural production but soon broadened to encourage investment by firms, households, and local governments. These measures were introduced incrementally and involved decentralizing authority over capital spending. A favored approach was for the central authorities to experiment with new policies in selected provinces, prefectures, counties, and even firms. If the experiments worked, they were quickly replicated. If they did not, the costs of failure tended to be contained and limited. Reforms were occasionally reversed if the government believed that growth was not being served or stability was being jeopardized.

Pragmatism and incrementalism were also behind the government's evolving objectives. When the reforms began in 1978, it is not clear whether the authorities had reached a consensus on any final objectives other than high growth. New reform objectives emerged as old ones were achieved. In the first few years of reform

FIGURE 1.9

Regional disparities account for a small share of inequality, 1985–95

Theil index

0.40
0.35
0.30
0.25
0.20
0.15
0.10
0.05
0
Actual inequality
Contribution of regional disparities
1985
1986
1987
1988
1989
1990
1991
1992
1993
1994
1995

Source: World Bank staff estimates.

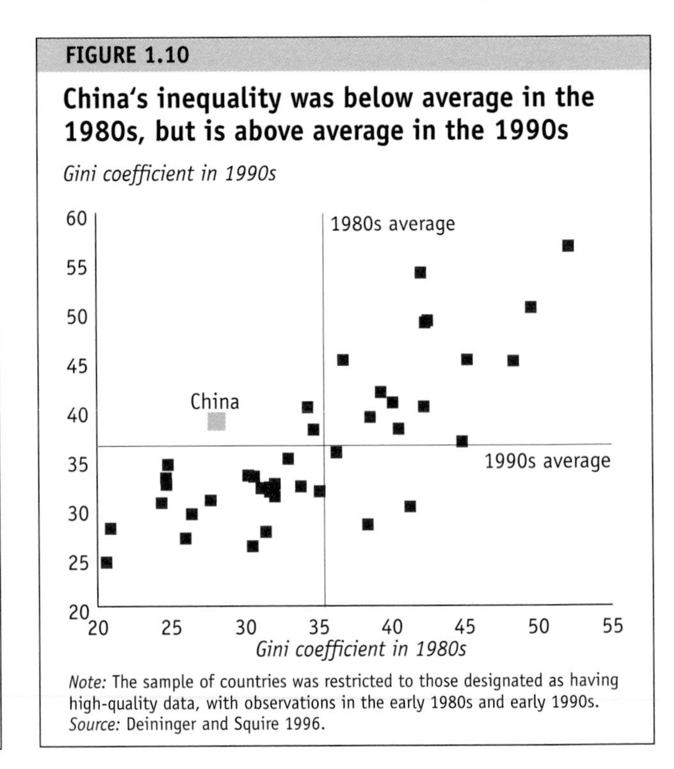

FIGURE 1.10

China's inequality was below average in the 1980s, but is above average in the 1990s

Note: The sample of countries was restricted to those designated as having high-quality data, with observations in the early 1980s and early 1990s.
Source: Deininger and Squire 1996.

BOX 1.3

The evolving objectives of reform

1978–79	A planned economy that is the "law of market exchange value"
1979–84	A planned economy supplemented by market regulation
1984–87	A planned commodity economy
1987–89	An economy in which the state regulates the market and the market regulates enterprises
1989–91	An economy with organic integration of a planned economy and market regulation
1993–present	A socialist market economy with Chinese characteristics

Source: Cao, Fan, and Woo 1995.

the objectives were modest. With success, they became more ambitious. In 1979, for example, the government called for the development of "a planned economy supplemented by market regulation." By 1993 the goal had matured to the creation of a "socialist market economy with Chinese characteristics" (box 1.3).

These features of China's reforms—decentralization, incrementalism, and pragmatism—are best illustrated in four key areas where reforms did much to stimulate growth: agriculture, rural industry, trade, and state enterprises.

Agriculture

It is hardly surprising that China's policymakers focused their initial reform efforts on agriculture. In the decade before 1978 they had been repeatedly disappointed by the performance of communal agriculture. Grain yields, after rising rapidly during the 1960s, had slowed markedly despite increased investments in agriculture (figure 1.11). This slowdown, combined with a long tradition of concern for food self-sufficiency and social stability in rural areas, made agricultural reforms a natural starting point.

Agricultural reforms were launched with large increases in the procurement prices for grains. In addition, farmers were allowed to sell above-quota production at market prices, a practice that had already started in some communes.[13] The government lowered grain quotas, increased grain imports, loosened restrictions on private interprovincial trade, and introduced special programs to increase production of cotton.

FIGURE 1.11

Rising grain yields

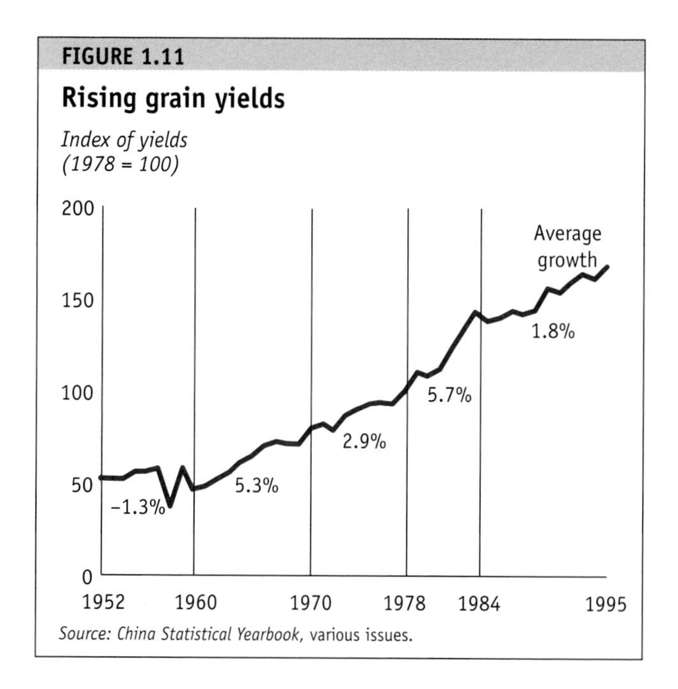

Source: China Statistical Yearbook, various issues.

But the most important feature of agricultural reform was the introduction of the household responsibility system, under which collectively owned land was assigned to households for up to fifteen years. The central government had banned the system, which transferred profits and production decisions from communes to households. But local governments encouraged it, and in late 1981 the central government signaled its approval (figure 1.12).[14] The household responsibility system went on to become the cornerstone of reforms in agriculture and arguably in the whole economy.

FIGURE 1.12

Expansion of the rural household responsibility system

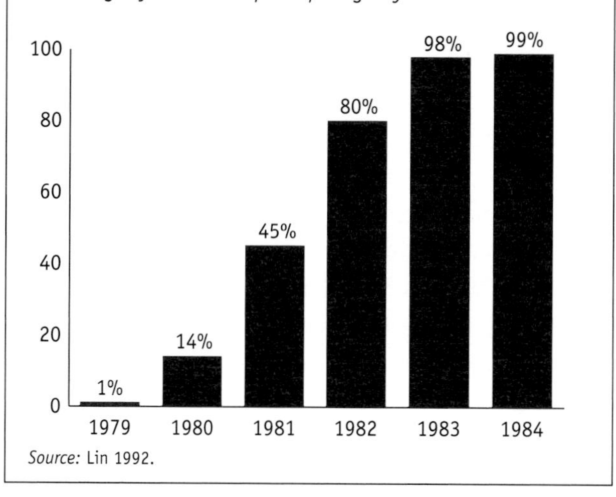

Source: Lin 1992.

Despite their partial and incremental nature, these steps proved very successful. Growth in agricultural yields accelerated, and the pattern of production changed as regions were able to specialize in crops of their choosing. By one estimate the introduction of the household responsibility system alone contributed nearly half of the growth in agricultural output between 1978 and 1984.[15]

Rural industry

The agricultural reforms initiated in 1978 were closely tied to the boom in rural industry that followed. Before 1978 rural industrialization had been an important feature of the commune-based system of agricultural development. But commune- and brigade-run enterprises did not produce consumer goods and were not geared to market demand. They were often simply workshops for agricultural machinery, designed to support agricultural production.

Agricultural reforms created the conditions for a rural industrial boom that continues to reverberate today. First, agricultural price increases and the production response that followed increased rural incomes substantially. As a result higher rural savings could be invested in rural industries, which generated higher returns than farming.[16] Second, the reforms boosted agricultural productivity, which freed surplus labor previously concealed in the commune system and so provided a steady supply of workers for rural industry. Third, higher rural incomes created a ready market for rural industry. Fourth, these new commercial activities were legitimized by changes in political philosophy and attitudes.

Official government support for these changes came only in 1984, six years after rural industrialization had begun. The government eased state controls on buying agricultural materials, making them available to rural enterprises for processing. Urban firms were encouraged to subcontract work to rural enterprises. More generally, production restrictions were relaxed, taxes were low, and enterprises were allowed to pursue whatever activity they chose.

The new policies helped maintain the momentum of growth of collectively owned enterprises through the 1980s. These enterprises accounted for 22 percent of industrial output in 1978. Their share reached 30 percent by 1984, without any overt central support, and climbed to 36 percent by 1988, where it has remained.[17] Individually and privately owned enterprises have become the driving force for industrial expansion as government policies toward these types of ownership have softened.

The great leap outward

The decentralization of authority evident in policies toward agriculture and rural industry was mirrored in China's economic relations with the world. By 1978 earning more foreign exchange had become a priority. Government foreign exchange reserves were negative, and the country faced a foreign exchange crisis.[18] It needed to import grains and fertilizer to support the agricultural reforms, and plant and equipment to close the widening technological gap with its neighbors.

China's economy opened to the outside world along three avenues. First was reform of the trade system. The government abandoned its reliance on a few foreign trade corporations, and within a few years thousands of corporations were allowed to trade internationally, many of them sanctioned by local, not central, government. Planning, too, was trimmed back. Until 1979 the state plan set procurement targets for 3,000 export products. By 1985 that number had shrunk to 100 (and today it is zero). The plan was replaced by import licensing, which was gradually relaxed as the foreign exchange constraint eased. Tariffs on imports were steadily lowered. Nontariff barriers, however, still affect a wide range of products, although they too have been reduced in scope and severity.

Second, to stimulate exports, the heavy hand of government in allocating foreign exchange gave way to a lighter touch. Exporters were allowed to retain a portion of their foreign exchange receipts, and individuals were gradually given more freedom to hold foreign exchange. The right to import was liberalized as foreign exchange markets developed with official support. Today's sophisticated interbank market for foreign exchange bears little resemblance to its earliest predecessor—the foreign exchange adjustment centers of 1985. Other banks were permitted to erode the Bank of China's monopoly on foreign exchange transactions. Perhaps most important, the government maintained a realistic exchange rate policy. It almost halved the exchange rate at the outset of the reforms and devalued the currency on four later occasions (figure 1.13).[19]

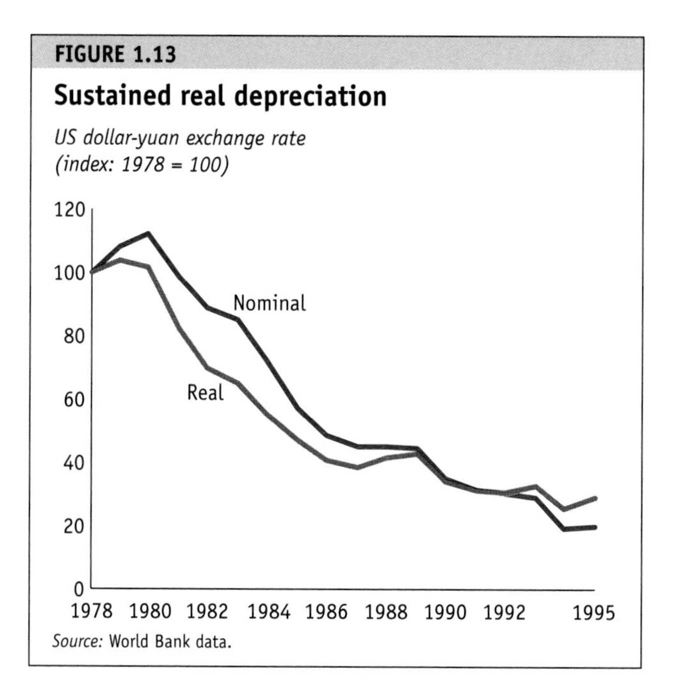

Third, the government gradually relaxed regulations on foreign direct investment. It established special economic zones that offered foreign investors special fiscal, infrastructural, and financial incentives. The first four special economic zones were created in 1980 and a fifth, in Hainan, was added a decade later. The zones' success in attracting foreign investment, stimulating trade, and invigorating growth soon led to demands from other towns and provinces for similar arrangements. By 1993 more than 9,000 economic zones had been established throughout China.[20] They came in all shapes and sizes—economic and technological development zones, hi-tech development zones, open coastal cities, export processing zones, free trade zones, a financial zone, and free ports.

Most of the zones were in coastal provinces and contributed significantly to the provinces' higher than average growth rates. One study found that the extra growth enjoyed by China's coastal cities in the late 1980s can be explained entirely by the extra foreign investment they were able to attract.[21] Foreign companies introduced advanced transport and electronics technology and helped upgrade key services, such as hotels. These new skills, together with China's ability to assimilate them, helped improve international competitiveness.[22]

State enterprises

The new policies put great pressure on state enterprises, which had been the mainstay of the economic system before 1978. Fed a diet of state subsidies for decades, many lacked the ability to adapt and innovate. With reforms, their dependence on public subsidies began to increase. The government's desire to contain its budget deficit meant that the state banking system became their main source of financial support.

Beginning in 1980 many enterprises acquired increasing autonomy over their operations. New freedoms included a slowly rising share of profits that could be retained for wage bonuses and new investment, greater autonomy over production decisions and wages, adoption of the "management responsibility system" (seen as a counterpart to agriculture's household responsibility system), and in some cases recruitment of new management (table 1.2). Central and local governments usually negotiated these new freedoms firm by firm. As a result the operational environment varied enormously between firms, across regions, and across sectors. In 1984 the Enterprise Bill of Rights for-

TABLE 1.2

Increased autonomy for state enterprises during the 1980s

(percent)

	1980	1981	1982	1983	1984	1985	1986	1987	1988	1989
Base retention rate[a]	7	19	22	30	34	37	39	38	39	39
Marginal retention rate[b]	11	12	11	14	17	17	19	23	26	27
Autonomy in production decisions[c]	7	8	10	14	25	35	40	53	64	67
Wage discretion[c]	1	1	1	2	5	9	12	20	32	35
Management responsibility system[c]	0	0	0	1	2	4	8	42	83	88
New management appointed after 1980[c]	9	9	15	25	40	40	61	75	85	94

Note: Based on a 1991 retrospective sample survey of state enterprises by the Chinese Academy of Social Sciences.
a. Portion of profits that could be retained if profits did not exceed a specified base level.
b. Portion of profits that could be retained if profits exceeded the base level.
c. Share of firms in the sample.
Source: Xu 1996.

malized these changes and created an additional impetus for growth.

The increased autonomy of state enterprises allowed them to benefit from China's dual-track pricing system, also introduced in the early 1980s. Under this system planners typically set a price for a commodity but allowed all above-plan output to be sold at market prices. Since the volume of planned output barely changed, enterprises sold more and more in the open market. Thus all growth and development occurred at market prices, which almost certainly improved resource allocation.[23] Because no enterprise was made worse off and some were clearly made better off, the reforms received enthusiastic support.[24] Today more than 95 percent of industrial output is sold at market prices.

The decentralization of management decisions boosted the productivity of firms.[25] But relative to the rest of the economy, state industrial enterprises languished, with slow growth and declining profits. In part this was because state enterprises, unlike their nonstate competitors, were required to provide job security and a range of social services, such as housing, education, and health care. Yet slackening performance also reflected a deeper malaise rooted in the poor investment decisions of the past and in an "iron rice bowl" system that did not penalize low productivity.

In the past several years, however, lower subsidies, tighter credit, and growing competition have begun to unmask the poor financial condition of many state enterprises, prompting new approaches to enterprise reform, especially at the local level. Examples include mergers, leases, corporatization, management contracts, worker and management buyouts, and bankruptcies. At the same time, the center is focusing on reforming and revitalizing 1,000 (from more than 100,000) state industrial firms that will form the core of China's modern enterprise system.

Initial conditions

Reforms in China spurred an economic response that continues to astound the world. High savings and structural change were undoubtedly helpful. But several features of the Chinese economy in 1978 made it particularly ripe for change.

First, China was simply ready for reform. The economic disruption of the Cultural Revolution, and before that of the Great Leap Forward, had undone many of the early economic benefits following the founding of the People's Republic. One study estimates that without the Great Leap Forward and the Cultural Revolution, output per capita in 1993 would have been double its actual level.[26] In 1978 real incomes in rural areas had been stagnant for more than a decade. The country was running short of foreign exchange to purchase essential imports. The widening technological gap between China and the rest of the world had become too large to ignore. Perhaps most important, China's neighbors in East Asia had demonstrated the potential for growth when high-savings economies adopted market principles.

Second, somewhat ironically, China enjoyed the "advantages" of backwardness (box 1.4).[27] More than two-thirds of the population lived in the countryside. They had no income guarantees and, despite rising

BOX 1.4

The "advantages" of backwardness

International experience shows that, allowing for differences in (for example) educational attainment and savings rates, poor countries tend to grow faster than rich ones. This is clear from the figure below, which plots the cross-country relationship between growth in 1960–90 against initial income, after controlling for differences in educational attainment, savings, and population growth. China's position is well above the regression line, indicating that its growth since 1978 has been unusually rapid, even for a very poor country.

China's growth, 1978–94: Benefiting from backwardness and more

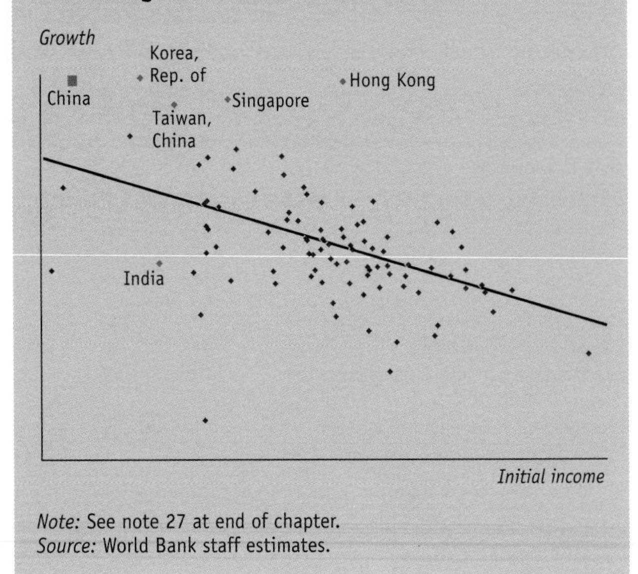

Note: See note 27 at end of chapter.
Source: World Bank staff estimates.

yields, for more than two decades their average incomes had barely grown. For them, the uncertainties of reform were less alarming than the difficulties of the present system. And agriculture's surplus labor meant that rural industry could achieve rapid, uninterrupted growth for almost two decades without facing wage pressures.

Third, contrary to popular perception, planning was less entrenched in China than it had been in other transition economies. In the 1970s central government agencies in the Soviet Union physically allocated about 60,000 different commodities through the plan. In China in 1978 the number was about 600, unchanged from 1965.[28] Not even the most determined planners could oversee an economy of the size and complexity of China's. Even at the height of planning, about 30,000 rural markets continued to operate, albeit with restrictions.[29] Smuggling was rife. So when commercial activities were legalized, Chinese entrepreneurs needed little encouragement to expand.

Fourth, China had always had a strong administrative capacity, especially at the provincial level. Over centuries China had developed a sophisticated system of local government to raise revenues and store grain as insurance against famine.[30] This tradition of local government became stronger under communist rule. So when reforms required administrative and financial decentralization, provincial governments were able to take on the new responsibility. Moreover, the central bureaucracy, severely weakened in the throes of the Cultural Revolution, quietly acquiesced to the shift in economic power away from the center.

Fifth, China had a skilled and disciplined labor force. Despite disruptions to education during the Cultural Revolution, literacy was high by the standards of most developing countries. The average worker had 3.6 years of primary education, almost half a year more than the developing country average of 3.2 years.[31] A relatively large portion also had a secondary education. The share of technicians and engineers in the industrial labor force was higher than in many newly industrializing economies of Southeast Asia.

Finally, the Chinese diaspora extended to virtually all corners of the world. Chinese minorities in several Southeast Asian countries had considerable economic power, and they figured prominently in the explosive growth of foreign direct investment in China. Over the past ten years there has been a wholesale movement of labor-intensive industries from Hong Kong and Taiwan, China, to the mainland, helped by a common language and a tradition of business through family connections. The Chinese diaspora brought more than money: it also contributed commercial expertise, knowledge of foreign markets, new approaches to management, new ideas on economic policies, and the latest labor-intensive technologies.

Notes

1. World Bank (1996).

2. The alternative GDP growth rate derived in box 1.1 is obtained by redeflating the expenditure components of GDP using alternative deflators, as follows. For consumption, we use the consumer price index (CPI). Although a national CPI was not available until 1986, one can be inferred from data on real and nominal per capita consumption growth reported in *China Statistical Yearbook 1996,* table 9.2. For investment, we use the price index for building materials in *China Statistical Yearbook 1996,* table 8.12. An alternative approach is to examine the deflators for the components of GDP by sector of origin, which leads to similar conclusions. For example, Woo (1995) suggests that the factory-gate price of industrial output is a better measure of price increases in industry. This approach results in a similar reduction in per capita GDP growth of about 1 percent. Finally, Summers and Heston, in their widely used database on per capita incomes at purchasing power parity, somewhat arbitrarily assume that consumption growth is overstated by 30 percent and investment growth by 40 percent, further lowering estimates of per capita GDP growth.

3. Maddison (1995). For China since 1950, Maddison uses the Summers-Heston GDP estimates, resulting in a GDP growth rate since 1978 that is substantially lower than official statistics.

4. Only two countries, Japan and Korea, have grown faster in the post–World War II era.

5. Nehru and Dhareshwar (1994). In only eight of the ninety-three countries covered in the Nehru-Dhareshwar data set do GDP growth rates exceed capital stock growth rates.

6. World Bank (1994).

7. See Yusuf (1994).

8. See Chenery and Syrquin (1975) for a classic reference on the international experience.

9. In some areas the predominant outflow of young male migrants has contributed to a "feminization" of agriculture, lowering the welfare of women who now have to toil on the farm and continue with their household responsibilities (especially child rearing).

10. Recent official statistics on the size of the nonstate, noncollective sector may be somewhat misleading because some corporatized state enterprises may be classified in this category.

11. Figure 1.8 is constructed as follows: Data on provincial per capita GDP and the nonagricultural share of GDP underlying the top figure are drawn from provincial statistical yearbooks. The other two figures use data on urban and rural incomes from China's household survey, adjusted to correct for various deficiencies as discussed in World Bank (1997). First, urban incomes are scaled up by a time-varying factor averaging 1.5, reflecting the value of in-kind incomes as estimated by the urban household survey team of the State Statistical Bureau and an estimate that urban costs of living are 15 percent higher than rural. Since in-kind incomes have an equalizing effect on income distribution (they account for a larger share of the income of poor

households), the urban Theil index is scaled down by a factor of 0.6. Rural incomes are adjusted by extending the results of Ravallion and Chen (1997) for four provinces to the entire country. Prior to 1990 rural incomes are scaled up by a factor rising from 1.06 to 1.14, reflecting the undervaluation of income from grain produced by rural households for personal consumption. Since this adjustment also has an equalizing effect, the rural Theil index is scaled down by a factor that falls from 0.86 to 0.69 in 1989. We assume that changes in the household survey since 1990 fully correct for the problems of rural grain undervaluation. The inequality measure is the Theil index, which can be decomposed into *rural-urban inequality,* which captures the variation of national average rural and urban per capita incomes around the national mean; *intraurban inequality,* which captures the variation of per capita incomes within urban areas; and *intrarural inequality,* which captures the variation of per capita incomes within rural areas.

12. World Bank (1997).

13. Quota prices for most agricultural products were increased by an average of 17.1 percent. Above-quota procurement commanded a 30–50 percent premium. The weighted average price increase was 22.1 percent (Lin 1992).

14. This discussion draws heavily on Lin (1992). Lin notes that nearly half of China's agricultural communes had been disbanded by the time the practice was authorized by the central government.

15. Lin (1992).

16. See Peng (1994) for a discussion of the reliance of township enterprises on local rural sources of finance.

17. *China Statistical Yearbook,* various issues.

18. Wall, Boke, and Yin (1993, table 12.1, p. 102).

19. Lardy (1992).

20. Wall, Boke, and Yin (1993, p. 5).

21. Wei (1993).

22. Yusuf (1994).

23. Naughton (1995, p. 8).

24. Lau, Qian, and Roland (1996).

25. Xu (1996).

26. Chow and Kwan (1996).

27. The figure in box 1.4 plots the partial correlation between initial per capita income and subsequent growth based on the following standard cross-country growth regression:

$$g = .039 - .012\ln(y_0) + .012\ln(i/y) - .022(n + \delta + x) + .008\ln(sec)$$
$$(.037)\ (.002)\ (.003)\ (.014)\ (.002)$$

where g is average annual per capita GDP growth between 1960 and 1990, y^0 is per capita GDP in 1960 (1978 for China), i/y is the period average investment/GDP ratio, n is the population growth rate, $\delta + x$ is the rate of depreciation plus productivity growth set equal to 0.5, and *sec* is the secondary school enrollment rate in 1960. Numbers in parentheses are standard errors. The sample covers ninety-two countries. China's growth is for 1978–94; initial income is for 1978.

28. Naughton (1995, p. 42). Qian and Xu (1992) argue that the Soviet Union included 25 million commodities in its economic plans, while China included about 1,200.

29. Naughton (1995, p. 38).

30. Will and Wong (1991).

31. Nehru, Swanson, and Dubey (1995).

References

Cai, F. 1996. "Long-term Urbanization and Key Urban Policy Issues in China." Background paper prepared for this report, with financial support from the Canadian International Development Agency.

Cao, Y.Z., G. Fan, and W.T. Woo. 1995. "Chinese Economic Reforms: Past Successes and Future Challenges." Paper presented at the meeting of the Asia Foundation Project Economies in Transition, Ho Chi Minh City, Hanoi, Beijing, and Ulan Bator.

Chenery, H., and M. Syrquin. 1975. *Patterns of Development, 1950–70.* New York: Oxford University Press.

China State Statistical Bureau. Various years. *China Statistical Yearbook.* Beijing.

Chow, G. 1993. "Capital Formation and Economic Growth in China." *Quarterly Journal of Economics* 108(3): 809–42.

Chow, G., and Y. Kwan. 1996. "Estimating Economic Effects of the Political Movements in China." *Journal of Comparative Economics* 23(2): 192–208.

Deininger, K., and L. Squire. 1996. "A New Data Set Measuring Income Inequality." *The World Bank Economic Review* 10(3):565–91.

Lardy, N.L. 1992. *Foreign Trade and Economic Reform in China.* New York: Cambridge University Press.

Lau, L., Y. Qian, and G. Roland. 1996. "Pareto-Improving Reforms through Dual-Track Liberalization." Stanford University, Department of Economics, Stanford, Calif.

Lin, J.Y. 1992. "Rural Reforms and Agricultural Growth in China." *American Economic Review* 82(1):34–51.

Maddison, A. 1995. *Monitoring the World Economy.* Paris: Organisation for Economic Co-operation and Development.

Naughton, B. 1995. *Growing Out of the Plan: Chinese Economic Reform, 1978–93.* New York: Cambridge University Press.

Nehru, V., and A. Dhareshwar. 1994. "TFP Growth in Industrial and Developing Countries." Policy Research Working Paper 1313. World Bank, International Economics Department, Washington, D.C.

Nehru, V., E. Swanson, and A. Dubey. 1995. "A New Database on Human Capital Stock: Sources, Methodology and Results." *Journal of Development Economics* 46(3):379–401.

Peng, Y. 1994. "Capital Formation in Rural Enterprises." In C. Findlay, A. Watson, and H. Wu, eds., *Rural Enterprises in China.* New York: St. Martin's Press.

Qian, Y., and C. Xu. 1992. "Why China's Economic Reforms Differ: The M-Form Hierarchy and Entry/Expansion of the Non-State Sector." *The Economics of Transition* 1(2):135–70.

Ravallion, M., and S. Chen. 1997. "When Economic Reform Is Faster Than Statistical Reform: Measuring and Explaining the Inequality in Rural China." World Bank, Policy Research Department, Washington, D.C.

Summers, R., and A. Heston. 1991. "Penn World Tables (Mark 5): An Expanded Set of International Comparisons, 1950–1988." *Quarterly Journal of Economics* 106(2): 327–68.

Wall, D., J. Boke, and X. Yin. 1993. *China's Opening Door.* London: Royal Institute of International Affairs.

Wei, S.J. 1993. "Open Door Policy and China's Rapid Growth: Evidence from City Level Data." NBER Working Paper 4602. National Bureau of Economic Research, Cambridge, Mass.

Will, P., and R.B. Wong. 1991. *Nourish the People: The State Civilian Granary System in China, 1650–1850.* Ann Arbor: University of Michigan.

Woo, W.T. 1995. "Chinese Economic Growth: Sources and Prospects." University of California at Davis, Department of Economics.

World Bank. 1991. *World Development Report 1991: The Challenges of Development.* New York: Oxford University Press.

———. 1994. "China: GNP Per Capita." Report 13580-CHA.

China and Mongolia Department, Washington, D.C.
———. 1996. *Poverty in China: What Do the Numbers Say?* Washington, D.C.
———. 1997. *Sharing Rising Incomes: Disparities in China.* Washington, D.C.
Xu, L.C. 1996. "The Productivity Effects of Decentralizing Cash Flow Rights and Management: A Theoretical and Empirical Analysis of the Chinese Industrial Reforms." World Bank, Washington, D.C.
Yusuf, S. 1994. "China's Macroeconomic Performance and Management During Transition." *Journal of Economic Perspectives* 8(2):71–92.

chapter two

Divining the Future

The strength of the Chinese economy over the past two decades does not guarantee that China will continue to grow rapidly in the future. After all, for most countries past growth is a poor predictor of future performance.[1] Even the Asian "tigers," whose growth has been spectacular for many years, show signs of fading as their economies mature.

China has some real advantages, however. Its high savings rate, relative stability, large domestic market, and record of reform bode well for future growth. Even so, the future will be challenging—probably increasingly so. Moreover, rapid economic growth is not the government's sole economic objective nor should it be. Growth must also be sustainable and its benefits must reach the poor and the vulnerable. And growth should protect the environment for future generations and ensure women their equal place in society.

These concerns are reflected in China's Ninth Five-Year Plan and its Fifteen-Year Perspective Plan (box 2.1). The two plans focus not just on the future pace of growth but also on the entire direction of development. They differ from previous plans because they focus on strategies, policies, and programs and place less emphasis on physical and quantifiable targets. Together, the plans lay out the government's approach to achieving rapid and sustained growth well into the twenty-first century.

These twin concerns—the pace and sustainability of China's growth—are also the themes of this chapter and indeed this entire book. This chapter also examines China's growth potential over the long term using a simple model of growth and structural change. It does not try to predict China's future, but it does illustrate what China could achieve if it stays the course.

Growth challenges

The various forces playing on China's growth rate are summarized in figure 2.1. The strengths and advantages have already been analyzed; but they are opposed by a formidable array of risks and challenges.

- *Incomplete foundations.* China is midstream in its transition to a market economy. The government's role in laying the institutional, social, physical, and legal foundations for market development will be crucial in completing the transition. To accomplish this task China needs an effective government, not a large one.

An effective government will create more room for markets to operate and distinguish between the roles of the market and those of the state. Indeed, a recurrent theme in the Ninth Five-Year Plan is the importance of firmly differentiating the basic role of the market (allocating resources) from that of the state (providing macroeconomic control and a policy framework in line with China's strategic needs and objectives).[2]

Pressing ahead with reforms to complete the transition to market is an immediate priority for China. This book therefore deals with it first (in chapter 3), laying out the issues and long-term agenda for developing the fiscal, financial, enterprise, and legal systems that will form the foundations for a strong market-based economy.

- *A deteriorating environment.* Achieving ambitious growth targets would be a hollow victory if it were not accompanied by improvements in people's physical environment. Along with the rapid economic growth, urbanization, and industrialization of the past two decades has come a dramatic increase in air and water pollution. In the future, needs for water and energy will have to be balanced by concerns for the environment. Rapid urbanization should be accompanied by programs that ensure clean air and water for city dwellers. Achieving these objectives will require a complex mix of regulations and market-based incentives.

BOX 2.1

China's Ninth Five-Year Plan and Fifteen-Year Perspective Plan

China's Ninth Five-Year Plan (1996–2000) and Fifteen-Year Perspective Plan outline the basic principles and priorities that will shape public policy in China well into the next century. According to the plans, rapid and balanced growth can continue only if China completes two fundamental transitions: from a traditional, planned economy to a socialist market economy and from extensive growth (based on increases in inputs) to intensive growth (driven by improvements in efficiency). In addition to calling for real GDP growth of 8 percent a year over the next five years, the plans lay out an ambitious agenda:

- *Maintaining the momentum of economic reforms.* The government is determined to contain inflationary pressures through budgetary and monetary restraint, supported by fiscal and financial sector reforms. It will concentrate on reforming the 1,000 largest state-owned enterprises and will pursue more intensive growth by promoting science and technology and developing five "pillar" industries: machinery, electronics, petrochemicals, automobiles, and construction.
- *Developing human resources.* The government has set itself the target of eliminating poverty by 2000, primarily through economic growth and targeted public expenditures. It also intends to make primary health care and nine years of compulsory education available to all and to expand coverage of social security benefits.
- *Strengthening agriculture.* The government aims to encourage basic agricultural products—especially grain, cotton, and oilseed—through land reform, agricultural research, improvements in rural infrastructure, and a range of other measures.
- *Protecting the environment.* Recognizing the environmental damage that has accompanied growth, the government intends to increase environmental regulation and enforcement, adopt environmental taxes, and increase public education and awareness.

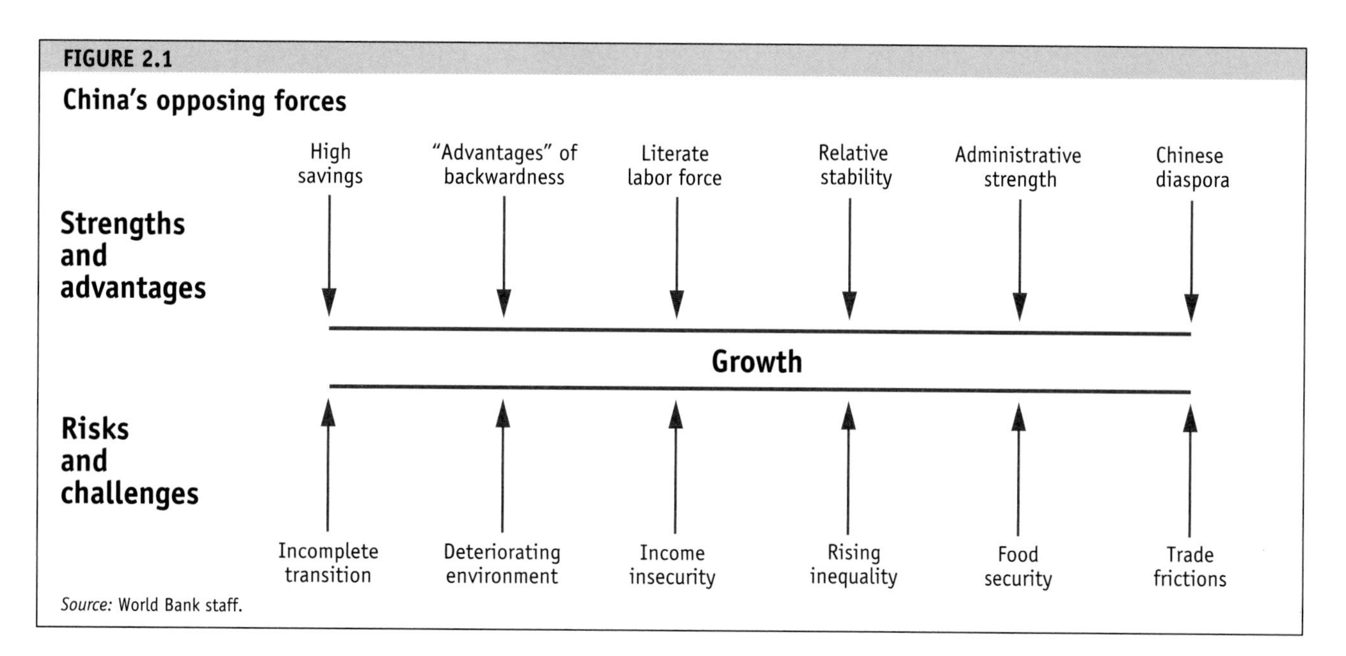

• *Income insecurity.* China has made great strides in introducing market forces into the economy. But the bracing effects of competition are being accompanied by increased risks, especially to employment and income. As China grows richer, it will require policies and institutions that ensure a caring yet competitive system to help the vulnerable manage these risks and to promote human potential in all dimensions.

This will require the creation of entirely new social structures for the twenty-first century. Changing employment patterns and foreseeable shifts in the age structure of China's population call for a fresh look at policies and institutions affecting labor markets, the welfare of the poor, the financial security of the elderly, and equal access for all (especially girls and women) to jobs, health care, and education.

• *Foodgrain security.* China's concerns about food security are as perennial as the harvest. Policymakers, facing a growing and prospering population and a difficult international environment, are acutely aware of these concerns. Indeed, food security tops the list of five priorities for "promoting sustained, rapid, and sound development of the national economy" in the Ninth Five-Year Plan and the Fifteen-Year Perspective Plan.[3]

Some of China's food security concerns are the inevitable consequences of geography. To be agriculturally self-sufficient, China must find a way to feed 20 percent of the world's population using just 7 percent of the world's arable land.[4] The government is exploring ways to achieve this objective—including stronger incentives for agricultural production and distribution, closer integration with world food markets, more rural infrastructure, greater efficiency in water use, and better flood control.

• *Trade frictions.* Although rapid export growth worked for many countries in the second half of the twentieth century, it may work less well for China in the twenty-first century. China is an enormous country with a potential capacity for more exports. Increasing exports, however, will require adjustments by both trade competitors and partners. Their governments may be more sensitive to trade issues today than they were in the 1950s. But the choice for China remains clear. Greater openness and transparency in its trade regime could enhance domestic production and allocative efficiency, facilitate the acquisition of new technology, and increase the flow of manufacturing and marketing information. And greater integration of China with the world economy would yield significant benefits for its major trading partners in East Asia, Europe, and the Americas.

Growth possibilities

China's Five-Year Plan for 1996–2000 targets GDP growth of 8 percent a year. In 1996, the first year of the plan, GDP growth reached 9.5 percent. To achieve the plan target by 2000, China will need to grow by an average of 7.6 percent a year over the next four years. By the standards of recent years, such growth does not seem unduly ambitious.

But what about the longer term? Could China keep growing so rapidly, not just for the next four years but for the next twenty-four? In particular, will high savings, structural change, and policy reform continue to be as powerful a combination as they have been since 1978?

Projecting long-term growth is not for the fainthearted. There is always the possibility of nonlinear change, as key variables start to behave quite differently. For China nonlinearities may be especially pronounced, given the rapid changes in the country's structure, the scale of its economy, and the underlying complexity of its growth dynamics. The World Bank has often been accused of overestimating the growth potential of its clients. In China, however, past Bank projections have tended to underpredict growth, often substantially (figure 2.2). In fact, China's GDP in 1995 was nearly double that predicted by the World Bank in 1985, the last time the Bank published a long-term study on the Chinese economy.

Nevertheless, projecting the future can be a useful exercise, if only to highlight the constraints that a particular path of savings and productivity growth place on long-term economic growth. The government is well aware of these constraints. Its Ninth Five-Year Plan emphasizes the imperative of more intensive growth, recognizing the fundamental importance of further productivity and efficiency gains for sustained, rapid development. Accordingly, this section uses a simple supply-side model of growth and structural change to examine China's growth potential over the next quarter-century.[5]

Even modest variations in savings and productivity growth have powerful effects, resulting in growth rates ranging from 4 percent to 8 percent a year (table 2.1). The cumulative effect of that gap amounts to more than a twofold difference in per capita incomes by 2020.

What do these alternative scenarios imply for policy? The links between the fundamental determinants of growth and the policies of government are many and complex. After all, the bulk of China's savings is the result of decisions by households and enterprises, not by the government. Similarly, many efficiency gains underpinning growth will be generated by individuals and firms pursuing their self-interest in competitive markets, rather than directly by actions of the state.

Even so, the government's actions are not irrelevant to long-term growth. The twin thrusts of the government's program—deepened reforms and continued stability—will profoundly shape the environment in which households and enterprises operate (figure 2.3). Successful reforms will increase efficiency gains by creating incentives for work and innovation and by establishing the essential foundations of the rule of law and secure property rights. A stable macroeconomic environment and financial system will enable savings to be channeled to their most productive uses. Finally, growth will be accompanied by further dramatic changes, especially in the structure of the labor force, that the government has a responsibility to manage in an orderly way. All this points to the importance of the government pressing ahead with efforts to deepen reforms and maintain stability in a large and rapidly changing economy.

To examine the prospects for growth more closely, consider a scenario in which China's savings rate declines from around 40 percent to a still-robust 35 per-

FIGURE 2.2

Trailing in the wind: World Bank projections of China's GDP compared with actual outcomes

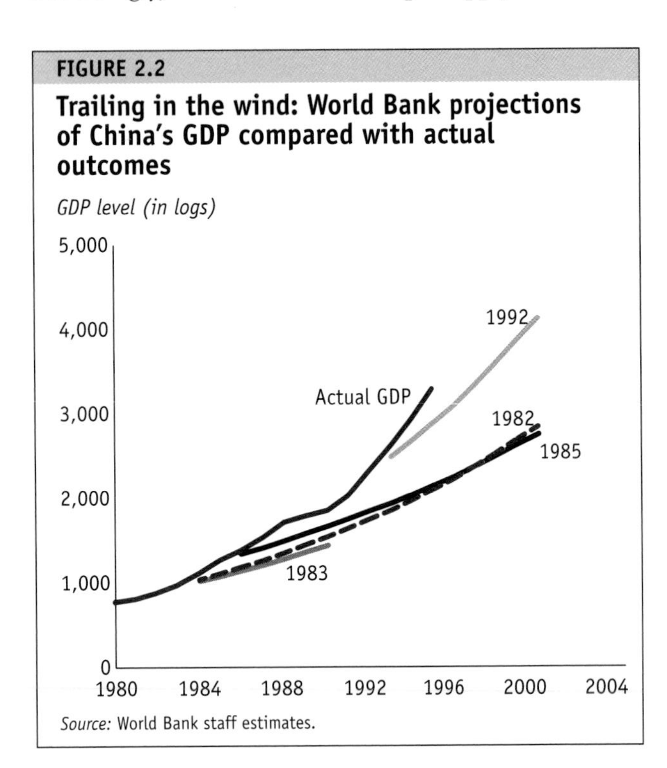

Source: World Bank staff estimates.

TABLE 2.1

GDP growth, 1995–2020: Alternative scenarios
(annual percentage growth)

	Total factor productivity growth		
Savings/GDP ratio	1	1.5	2
20	4.2	4.9	5.5
25	4.8	5.5	6.4
30	5.4	6.1	7.2
35	5.9	6.6	7.6
40	6.4	7.2	7.9

Note: Shaded area indicates likely growth rates.
Source: World Bank staff estimates.

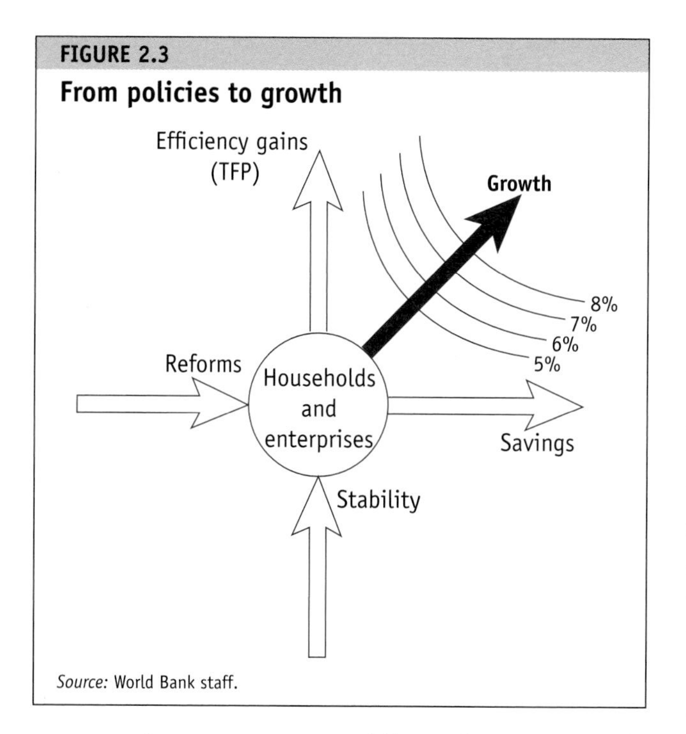

FIGURE 2.3

From policies to growth

Source: World Bank staff.

cent over the next ten years, while productivity growth fades slightly, to a more modest 1.5 percent a year. In this scenario GDP growth would be 8.4 percent a year between 1996 and 2000 (comfortably in line with the Ninth Five-Year Plan target) and would average 6.6 percent over the twenty-five years until 2020. By 2020 per capita incomes in China would be approaching those of Portugal today but would still be less than half those in the United States (figure 2.4).[6]

As the numbers suggest, the pace of GDP growth will slow over time, from 9–10 percent today to 5 percent in 2020 (table 2.2). Three forces are expected to contribute to this decline. The first is demographics. Projections indicate that by 2020 the labor force will essentially have stopped growing as a result of slower population growth and a changing age structure. Second, just 10 percent of China's expected capital stock in 2020 is in place today. As capital accumulates, each additional unit can be expected to contribute less to output—the economic law of diminishing returns. Third, as the economy matures, structural change will provide a smaller boost to growth because resources in the economy will be more efficiently allocated. In particular, China will reap fewer benefits from transferring surplus labor out of agriculture and from one-shot efficiency gains. And as China narrows the technology gap with other economies, its impetus for technical progress will ease.

From a sectoral perspective the decline in projected GDP growth is partly a consequence of lower industrial growth (see table 2.2). Here, much will depend on the pace of reforms. If the government falters in its commitment to a dynamic nonstate sector (which provided the foundation for recent growth) and slows reform of state enterprises, the drop in industrial growth may be even steeper. Even so, industry will continue to account for a much larger share of GDP than is the norm for low- and middle-income countries—the legacy of a planning system that emphasized rapid industrial growth above all else (figure 2.5).

Agriculture is expected to grow at about the same rate as it has in the past. But even keeping up this pace will be a major challenge (see chapter 5). The agricultural commitments in the Ninth Five-Year Plan could boost the sector's growth in the short term, but some slowdown is likely thereafter.

Services are expected to be by far the most dynamic sector of the economy. Despite strong growth in the

FIGURE 2.4

A long way to go: China's per capita GDP in 2020

1994 U.S. dollars

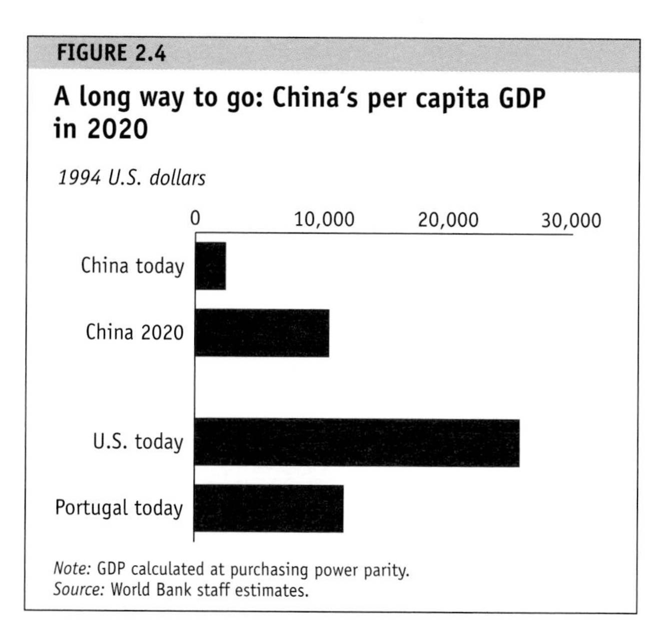

Note: GDP calculated at purchasing power parity.
Source: World Bank staff estimates.

TABLE 2.2

Sectoral underpinnings to GDP growth
(percent)

	Actual	Projected			
	1985–95	1996–2000	2001–10	2011–20	1995–2020
GDP	9.8	8.4	6.9	5.5	6.6
Agriculture	4.2	3.1	4.2	3.7	3.8
Industry	13.1	9.2	6.6	5.4	6.6
Services	9.8	9.7	8.1	6.0	7.6

Source: World Bank staff estimates; *China Statistical Yearbook 1996.*

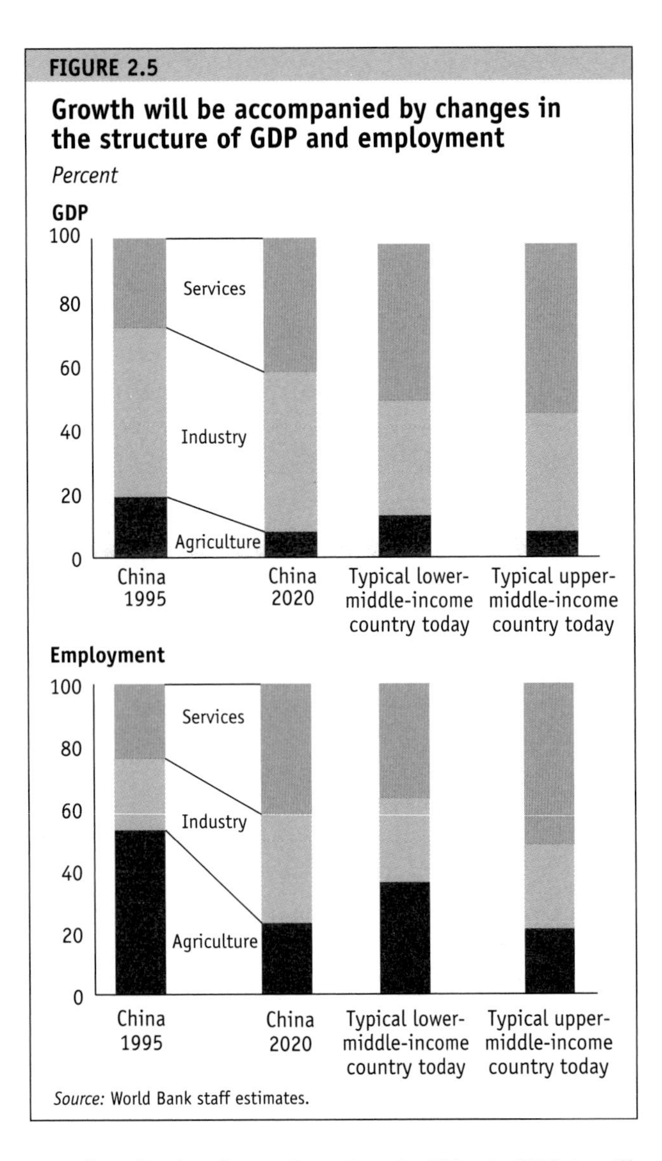

Source: World Bank staff estimates.

past decade, the share of services in China's GDP is still well below that of a typical low-income (let alone middle-income) country. This share could change dramatically over the next twenty-five years as China acquires the characteristics of middle-income market economies.

A key ingredient in this growth scenario is the transformation in the composition of the labor force. Agricultural employment is expected to fall from more than half of total employment today to one-quarter in 2020, a share comparable to that in upper-middle-income economies. The corresponding increase in the nonagricultural labor force will be felt in services, whose share in total employment will rise by 18 percentage points, as well as in industry, where the continued shift from state to nonstate enterprises will increase the labor intensity of production.[7]

In human terms this transformation of the labor force is staggering: for agriculture alone it will fundamentally change the lives of 200 million people. Managing this transition, and the inevitable pressures it will place on urban infrastructure, labor market institutions, and social programs, is one of the biggest challenges the government faces. But the transformation should be viewed as an opportunity, not a threat. China's surplus labor is one of its greatest resources. Efficiently handled, it will provide a powerful impetus to growth.

These scenarios paint a complex picture of China's future. Rapid growth over the next quarter-century is clearly possible. But achieving needed savings rates and productivity growth will require maintaining the momentum of reforms and skilled macroeconomic and sectoral management. These themes of supportive reform and successful management of change are woven throughout this book.

Notes

1. Easterly and others (1993).
2. Li (1996, p. 10 in the English translation).
3. Li (1996, p. 10). The remaining priorities are adjustments in industrial structure, development of regional economies, macroeconomic stability, and rising living standards.
4. *China Statistical Yearbook 1996.* Arable land figures refer to total cropland in 1995, which was about 95 million hectares, or 7 percent of the world's 1,450 million hectares.
5. See annex 2 for a complete description of the model.
6. This comparison is based on the standard practice of applying real (projected) local currency growth rates to current World Bank estimates of per capita GDP at purchasing power parity (PPP). These projections are subject to the same substantial margins of error as are estimates of China's current level of per capia GDP at PPP.
7. Hu (1996).

References

China State Statistical Bureau. Various years. *China Statistical Yearbook.* Beijing.

Easterly, W., M. Kremer, L. Pritchett, and L. Summers. 1993. "Good Policy or Good Luck? Country Growth Performance and Temporary Shocks." *Journal of Monetary Economics* 32: 459–83.

Hu, A. 1996. "Employment—Number 2 Task of China's Development." Background paper prepared for this report, with financial support from the Canadian International Development Agency.

Li, Peng. 1996. "Report on the Outline of the Ninth Five-Year Plan for National Economic and Social Development and the Long-Range Objectives to the Year 2010." Speech delivered at the Fourth Session of the Eighth National People's Congress, March 5, Beijing.

chapter three

Laying the Foundations

The past eighteen years in China have underscored the importance of markets for growth and prosperity. But the country's transition from a command to a market economy is incomplete. Many parts of the Chinese economy still have an undesirable blurring of governmental and commercial functions, a holdover from the days of central planning. These poorly defined roles create conflicts of interest within government and ultimately could harm the economy. This chapter argues that further separating the roles of government and markets and clarifying rights and responsibilities will help to lay the foundations for sustained rapid growth and improve the quality of people's lives.

Good markets start with good government. This is true because effective governments serve markets rather than make markets serve government. The primary instrument of an effective government is its budget. Strong government finances anchor macroeconomic stability and fund a range of essential services that markets may fail to deliver. In China strong government finances are also needed in the short term to underwrite costly reforms, and in the medium term to build the social, physical, and institutional infrastructure for continued rapid growth. In the near term achieving these objectives will require more government revenue. Over the longer term it will require building fiscal institutions that balance planners' priorities with people's preferences.

Next, the government must leave more room for markets to operate. Despite nearly two decades of reform, state and market remain intertwined. For example, state enterprises still provide a wide range of social services to workers and their families. Similarly, the four largest state commercial banks direct their lending in support of the government's policy objectives. Transforming state enterprises and commercial banks into vibrant, market-oriented institutions, and so to nurture emerging private economic activity, is probably the biggest challenge the government faces in the near term.

But a diminishing role for state enterprises and state commercial banks does not mean that the government should abdicate all responsibility for the development of markets. On the contrary, it implies a changing role for government as it reorients itself toward key services, sound institutions, and credible policies in support of markets. These include such fundamentals as establishing the rule of law, providing the institutional framework that guarantees a stable economic environment, and intervening in areas where markets fail to function adequately.

In all three areas—restoring and reshaping government finances, making room for markets, and supporting the development of markets—the government is now midway through reforms. The close links between the three make progress slow and often laborious. Nonetheless, the overwhelming impression is one of relentless forward movement.

Restoring and reshaping government finances

Government finances have been weakening for several years. The shortage of budgetary revenues has already hampered spending on essential services with long-term benefits for growth and the quality of life. Budgetary difficulties have generated inflationary pressures and imposed additional strains on the financial sector. This situation cannot continue forever. Eventually, weak government finances will debilitate the economy.

These fiscal pressures raise four long-term issues. First is the need to set priorities to ensure the best use of scarce government resources. Second is the imperative of improving tax collection, including a search for new types of taxation. Third is the urgency of establishing an effective mechanism of intergovernmental grants to redistribute resources from richer to poorer provinces. Finally, there is a need to develop institutions that will make budget formulation more transparent to the public and responsive to its needs.

Reshaping government expenditures

Between 1978 and 1995 budgetary revenues tumbled from 35 percent of GDP to 11 percent. Of this 24 percentage point drop, about 15 percentage points were directly attributable to lower tax contributions from industrial state enterprises (figure 3.1),[1] as price reforms and more intense competition lowered their profits. This decline in revenue, combined with the government's desire to keep the budget deficit modest, has brought a similar decline in government expenditures as a share of GDP. Today, government expenditures are about 12 percent of GDP, well below the developing country average of 32 percent.[2] Investment has suffered the most, falling from 16 percent of GDP in 1978 to less than 3 percent in 1995 (see figure 3.1).

At first glance these trends seem consistent with China's move to a market economy. But such impressions are deceptive. The reduction in direct financing of investment has been partly offset by increases in extrabudgetary spending and government-directed bank credits. Together these off-budget and government-directed resources are more than five times as large as the investment financed through the budget (figure 3.2).

Some extrabudgetary spending is financed from quasi-tax revenues raised through formal levies, such as those on electricity, vehicles, and railroad freight.[3] Some is funded by extraordinary levies imposed, sometimes arbitrarily, by provincial authorities. The rest represent the profits of state-owned enterprises controlled by local governments. There are few mechanisms to mon-

FIGURE 3.1

Reshaping government expenditures

Budgetary revenues have dwindled . . .

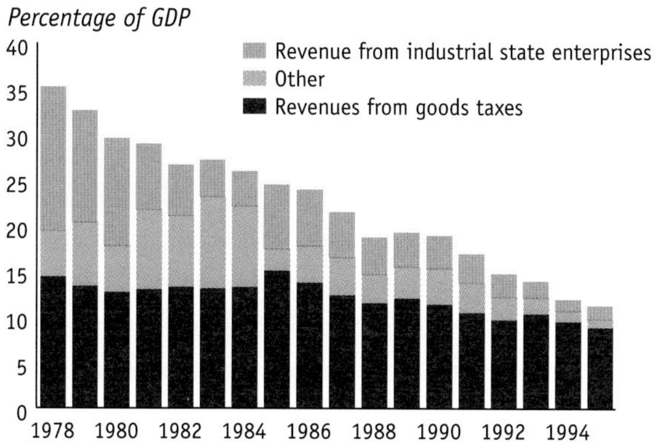

Percentage of GDP
40
35
30
25
20
15
10
5
0
Revenue from industrial state enterprises
Other
Revenues from goods taxes
1978 1980 1982 1984 1986 1988 1990 1992 1994

. . . and expenditures have followed

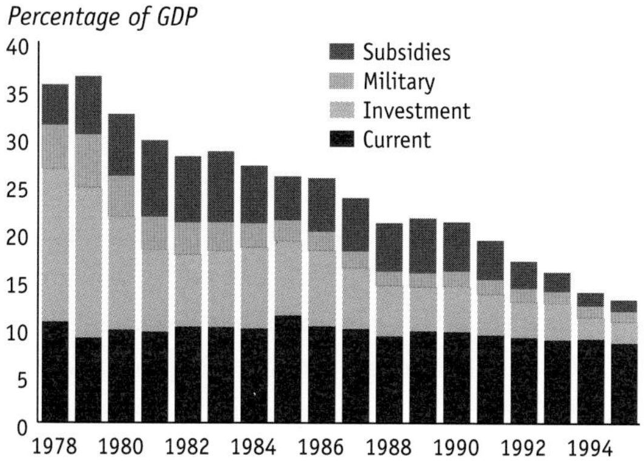

Percentage of GDP
40
35
30
25
20
15
10
5
0
Subsidies
Military
Investment
Current
1978 1980 1982 1984 1986 1988 1990 1992 1994

Source: World Bank staff estimates using *China Statistical Yearbook,* various issues.

FIGURE 3.2

The budget finances a small share of government-controlled investment

Extra-budgetary fiscal resources 45%
Central 5%
County 4%
Foreign 4%
Bank loans 39%
District/city 7%
Provinces 10%
Other 15%
Budget 15%

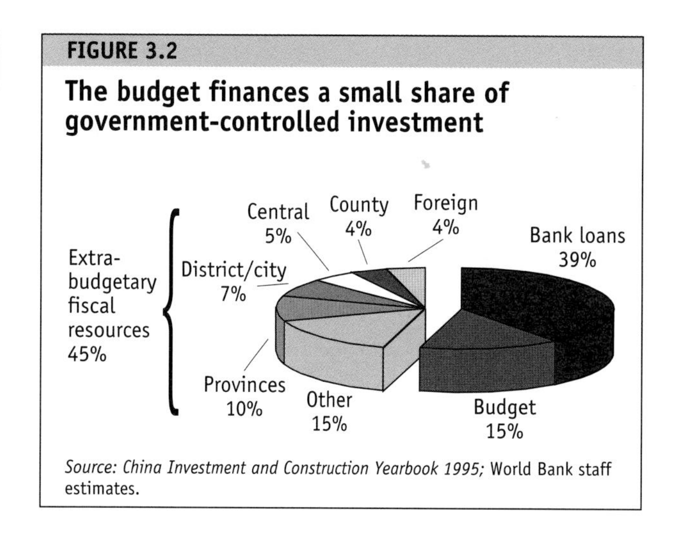

Source: China Investment and Construction Yearbook 1995; World Bank staff estimates.

itor these funds or the effectiveness with which they are invested.

One undesirable result is that the system discriminates against projects that may be socially and economically important but financially unprofitable. Thus spending on health, education, poverty alleviation, pensions, infrastructure and the environment has not kept pace with needs (box 3.1). It is neither a healthy nor a hopeful sign when schools and hospitals are forced to scramble for extrabudgetary funds to pay their bills. Such conditions inevitably undermine the quality and quantity of public goods and services, with undesirable implications for rapid, high-quality growth.

The overall expenditure shortfall in these high-priority areas has been estimated at 4.6 percent of GDP (table 3.1).[4] This is a conservative figure, however, because it does not include the possibility that services will cost more in the future.[5] Nor does it include possible additional expenditures in support of future economic reforms—spending on such items as unemployment benefits, redundancy payments, monetized housing benefits, or the costs of banking reform.

This pattern of public expenditures—too much government-controlled investment in state enterprises and too little government support for high-priority areas—is inconsistent with China's move to a market economy and inimical to sustained, rapid growth. Altering this pattern is essential. Many options are possible, but the basic features would be the same: devoting additional resources to social programs and the environment, adequately funding key reforms, and scaling back government control of investments.

BOX 3.1

Unhealthy trends in public outlays

Budgetary expenditures on health (excluding outlays for the government insurance program covering civil servants) fell from 0.8 percent of GDP in the 1980s to 0.5 percent in 1993. But the actual decline was much greater. In the years before reform, rural collectives were largely responsible for health expenditures and services in rural areas; most of these expenditures were outside the budget. With the abolition of rural collectives, rural communities lost the ability to finance local government functions, especially basic health and education. In 1981, when rural collectives still existed, 71 percent of the Chinese population had access to government health facilities. By 1993 this figure had shrunk to 21 percent, mostly in urban areas.

Source: World Bank 1997b.

TABLE 3.1
Proposed additional expenditures and financing, 1994

Expenditure	Percentage of GDP
Actual budgetary expenditures	14.1
Extrabudgetary expenditures	3.8
Proposed additional expenditures	4.6
Education	0.9
Health	1.4
Poverty alleviation	0.2
Environment	0.2
Infrastructure	1.0
Social insurance	1.0
Total proposed expenditures	*22.5*
Financed by	
Budgetary revenues	12.4
Extrabudgetary revenues	4.1
Additional revenue required	*6.0*

Source: World Bank 1996b.

Reducing its role in capital spending will allow the government to focus on a narrower range of interventions that would contribute most to economic development. Outside this narrower range, capital spending would be financed through the internal funds of enterprises, the banking system, or capital markets, based on standard commercial criteria. This approach would help banks and enterprises commercialize faster, strengthen their financial position, and improve their competitiveness in domestic and international markets.

Over the next several years, however, government-directed investments will continue to exceed budgetary resources. Like reforms in other sectors, adjustments may have to be made gradually. As the government redefines its role in the economy, arrangements will be needed to smooth the transition.

China's policy banks have been created for precisely this purpose. They are essentially fiscal institutions created to free the banking system from the yoke of government-directed investments. In a year or two the government expects all government-directed investments outside the budget to be financed through the policy banks.

But this should be just an interim goal. A longer-term goal could be to fold all government investments into the budget. This could be achieved by 2010, certainly by 2020. It will take debate and discipline to restrict government investments to high-priority areas. It will also mean boosting government revenues to levels more consistent with China's stage of development and potential for growth.

Restoring revenues

After several years of decline, tax revenues as a share of GDP stabilized in 1996, following wide-ranging tax reforms two years earlier. Now the authorities face the even harder task of reversing the earlier decline. Achieving this goal will require better tax administration, a broad tax base, and new taxes that combine social, economic, and revenue objectives. In the short run improving compliance would likely do the most to boost revenues; phasing out the plethora of tax exemptions for domestic and foreign firms would also help. In the long run the government could consider further tax policy changes in support of reforms in environmental and social policy and could do a better job of taxing nonstate enterprises.

Compliance can be improved in three areas in particular. First, China has the capacity to raise the compliance rate for the value added tax (VAT), which accounts for almost half of government revenues, from roughly 70 percent today[6] to 85 percent by 2000 and to 90–95 percent by 2010, putting it in the same league as the top tax performers among developing countries. Second, by 2000 China could double revenues (as a share of GDP) from personal income taxes by keeping the exemption level constant in nominal terms. Third, the government could merge domestic and foreign corporate income tax rates, a change now under consideration. Special rates for foreign enterprises are no longer needed to attract foreign investors and should be eliminated.

In the medium term the government could combine its reform and revenue objectives. For example, a pollution tax would discourage the use of coal and petroleum and encourage the development of pollution-reducing technologies. Similarly, some form of social security contributions will be needed to finance the basic pension pillar for (urban) workers. International experience suggests that payroll contributions for a generally accepted purpose such as pensions tend to meet little taxpayer resistance.

Over the long term doing a better job of taxing nonstate enterprises is an extremely high priority. They are, after all, the fastest growing segment of the economy.

But many of them keep poor accounts (if they keep them at all), so it will take some innovative approaches to tax them fairly. Other longer-term measures to consider include taxpayer self-assessment, efficient audit procedures, computerized information systems, reorganization of tax administration, improved incentives for tax collection, close cooperation of national and local tax services, improvement in presumptive taxation methods, and implementation of a streamlined system of stiff but fair monetary fines for tax offenses.

Together these measures could raise incremental revenues equivalent to 6 percent of GDP by 2000 and an additional 4 percent of GDP by 2020 (table 3.2). These measures would bring total government revenues—inclusive of extrabudgetary revenues—to about 24 percent of GDP. But achieving even this relatively modest ratio will require a monumental effort by the tax authorities.

Expanding intergovernmental transfers

There will always be a revenue gap between China's poorer and richer provinces. Without an effective system for transferring revenues, underlying inequalities will be reinforced by unequal spending on health, education, environmental protection, and other public services.

The current system permits few fiscal transfers from richer to poorer provinces or counties. Each region is required to be more or less fiscally independent, tailoring its public expenditures to the revenues it can collect. As a result per capita expenditures vary widely across provinces, in line with per capita incomes (figure 3.3). Poorer provinces would obviously welcome transfers from the center. But richer ones would also benefit. Transfers would ease pressures for migration from poorer to richer provinces and strengthen the infrastructure of poorer provinces, which would help open markets in the interior and integrate it with the economic mainstream. The same benefits would flow from transferring resources to the poorest counties, much as happened with the government's 8–7 Plan, aimed at eliminating poverty by 2000 (see chapter 4).

TABLE 3.2

Estimated increase in revenues by 2000 from recommended tax measures

Measure	Additional expected revenue in 2000 (percentage of GDP)
Value added tax	1.1
Personal income tax	0.8
Corporate income tax	1.2
Pollution levies	1.0
Social security contribution	0.9
Improved tax administration	1.0
Total	6.0

Source: World Bank 1996b; World Bank staff estimates.

Building transparent, responsive fiscal institutions

The demands and priorities of government finances change over time, so it is important that fiscal institutions be able to react quickly to new situations. That kind of agility requires institutions that are transparent and responsive.

International experience shows that transparency in budgetary procedures affects overall fiscal performance. China's system for setting aggregate spending is far from transparent. As noted earlier, a significant portion of expenditures is financed from extrabudgetary sources and through the banking system.[7] The one-year planning horizon for public investments encourages

FIGURE 3.3

Per capita provincial government expenditures vary with provincial income

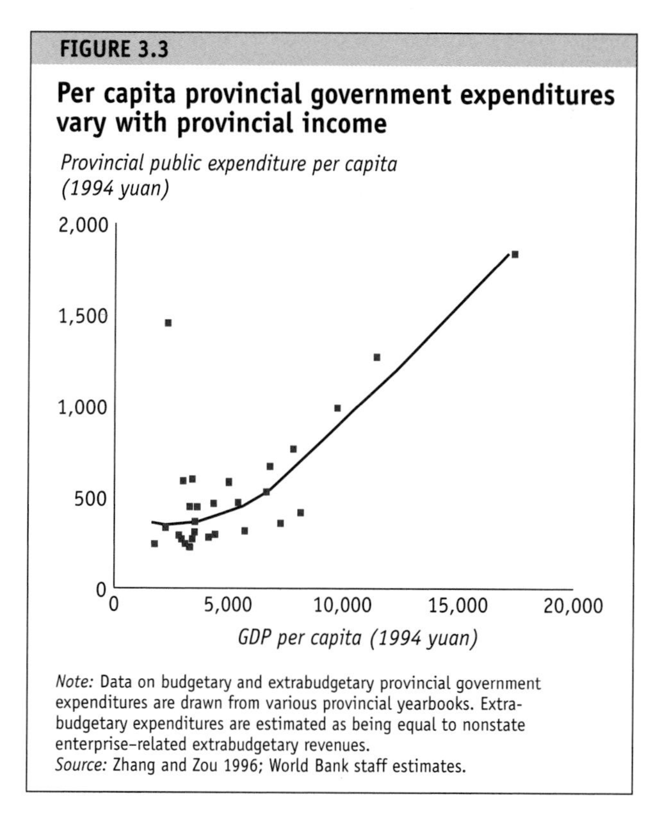

Note: Data on budgetary and extrabudgetary provincial government expenditures are drawn from various provincial yearbooks. Extrabudgetary expenditures are estimated as being equal to nonstate enterprise-related extrabudgetary revenues.
Source: Zhang and Zou 1996; World Bank staff estimates.

overinvestment by state-owned units, and banks find it hard to resist pressures to lend. As a result consolidated government deficits are much larger than the formal budget deficit.[8] Between 1986 and 1994 the consolidated deficit may have averaged from 4.9 to 5.7 percent of GDP—more than twice the 2.2 percent deficit in the official budget.[9]

The government recognizes these problems. It has been reviewing all extrabudgetary expenditures and has already incorporated quasi-tax revenues from electricity, vehicles, and railroad freight into the 1997 budget.[10] Furthermore, the state commercial banks should reduce and eventually eliminate the practice of financing government-directed investments. Another important step would be for the government to plan its expenditures, especially investments, within a medium-term economic framework. To discourage overinvestment, explicit rules could set spending limits and penalize agencies that overspend. To make the process more transparent, the government could publish figures comparing planned spending limits with actual expenditures.

Another area requiring improvement is the allocation of expenditures among priority uses. More should be done to build consensus on priorities and then to stick to them. The government already devotes considerable attention to developing a strategic outlook through its five-year plans and fifteen-year perspective plans. But the links between these plans and annual expenditure allocations are blurred.

To increase clarity, all government spending (including extrabudgetary items) should be incorporated in the budget, together with information on project details and tradeoffs between projects. It would also be useful to set out the rules for project selection, such as financial, economic, and social rates of return. Exposing government plans to wider public discussion, culminating in a debate in the National People's Congress, would help strike a balance between planners' priorities and people's preferences.

Making room for markets

The state still plays a big role in China's economy. State enterprises account for one-third of industrial output, two-thirds of urban employment, and more than half of investment in fixed assets.[11] And despite the entry of many new financial institutions in recent years, four large state commercial banks still account for two-thirds of the assets of the entire financial sector.

As industrial state enterprises have faced stiff domestic competition and liberalized market prices, their losses have mounted. The government has propped them up with subsidies and cheap credit, and the red ink of state enterprises has seeped into the portfolios of banks. Sluggish state enterprises erode the credibility of China's reforms and frustrate the development of markets in industry and finance.

There is little disagreement in China about the need to reform state enterprises and the financial system. The challenge is to introduce the reforms while maintaining rapid growth and economic stability. In the case of state enterprises the government is concerned about the consequences of reform for unemployment and social peace. In the financial sector the concern is with protecting the integrity of financial institutions and preserving depositor confidence while markets are given a bigger role.

Reforming enterprises

China has 305,000 state enterprises, 118,000 of which are industrial. Some are performing well, even in the global marketplace. But several indicators point to the generally poor performance of state industrial enterprises relative to nonstate firms. State firms have lagged in output, employment, and productivity growth. Their return on assets is estimated at just 6.0 percent, compared with 8.4 percent in collectives and 9.9 percent in joint ventures. This share of loss-makers has grown from 26 percent of the total in 1992 to 50 percent in 1996; as a group these weak performers reported operating losses in 1996 equivalent to 1 percent of GDP.[12] As a result the role of state enterprises in many industries has dwindled to the point where state ownership in China today resembles that in France in the 1980s (table 3.3).

The agenda for reforming state enterprises is long and difficult. It is still hard to obtain reliable financial information on their performance. They do not have adequate market-based autonomy and risk-reward incentives. The erosion of government authority over many enterprise decisions has provided opportunities for asset stripping and opportunistic behavior by managers, workers, and sometimes even government officials. Although there have been some recent bankruptcies, government (especially local government) is reluctant to shut down enterprises that have little hope of financial viability. Bankruptcy procedures are time-consuming and costly, and the government does not want to exacerbate urban unemployment.

But progress has been slow. These are, after all, complicated reforms. The most crucial is the transfer of pension, health, and education obligations from state enterprises to government bodies. True, some municipalities are pooling pension obligations across firms and earmarking payroll taxes for pension, unemployment, and health benefits. Some local authorities are taking over schools and hospitals previously run by enterprises. And some state enterprises are reducing housing subsidies by raising rents (and wages, to compensate). But these examples are the exception rather than the rule. And they have not fundamentally affected most enterprises.

The simple truth is that the authorities have neither the administrative capacity nor the appropriate information to effectively oversee the sprawling state enterprise empire. Absent effective oversight by their "owners," many state enterprises are controlled completely by their managers. Even in corporatized enterprises the same people often serve as senior executives of the firm and members of the board of directors. Under the right conditions such insider-dominated firms could still be efficient. But as other transition economies have found, such firms also harbor many risks—asset stripping, poor investment decisions, and decapitalization through excessive increases in wages and other private benefits (box 3.2). Enterprise assets could gradually be siphoned into private pockets, leaving the government to assume the liabilities.

Asset stripping and excessive wage compensation are widespread in Chinese state enterprises. A 1994 survey of 124,000 state enterprises found that asset losses and unaccountable expenses accounted for 11.6 percent of the assets of the firms sampled.[13] One estimate puts the annual loss of state assets at 30–100 billion yuan, or 2–9 percent of annual capital investment by state-owned units.[14] Half of the new limited liability companies established in Sichuan and Shanghai in the past few years are in the financial sector, which suggests the widespread creation of "shell" companies to drain assets.

This situation presents a dilemma for the Chinese authorities. Managers need autonomy to improve efficiency, but the government has neither the information nor the administrative capability to ensure that managers are pursuing that objective. So rather than reforming all state enterprises at once, the central authorities are focusing on 1,000 large state firms and on plans to invigorate them and make them the core of modern enterprise. About 880 of the firms are in the

TABLE 3.3

Sectoral composition of public ownership, China and France

(percent)

Industry group	China 1994[a]	France 1986[b]
Resource extraction	71	32
Utilities	65	86
Scale intensive		
Metallurgy	59	67
Chemicals	49	31
Transport machinery	51	46
Contestable[c]		
Food and textiles	43	1
Electronics	28	43
Other	22	10
All industry	40	29

a. State ownership in gross output.
b. Public enterprise share of sales.
c. Industries in which entry is relatively costless.
Source: China Statistical Yearbook 1995; Blanc and Brule 1993.

BOX 3.2

The privatization of assets and the socialization of costs: Enterprise reforms in Bulgaria

After the fall of the Berlin Wall, a coalition government in Bulgaria introduced sweeping reforms. Although swift privatization was part of the program, a change in administration in 1991 blocked this move for four years. During this period the Bulgarian government lost much of its capacity to monitor state enterprises, including the largest. Managers diverted assets and cash flow to themselves, leaving the enterprises with the liabilities. State enterprise losses mounted, reaching 12 percent of GDP in 1992–94. The losses were covered by bank loans, leading eventually to a banking crisis in 1995. By then banks' nonperforming assets had reached an astonishing 80 percent of all loans.

Source: World Bank 1997a.

industrial sector, accounting for two-thirds of the assets of industrial state enterprises and the majority of their sales, tax receipts, and profits.

A key element necessary for the invigoration of these 1,000 enterprises is an improvement in their governance, especially in two areas:

- Clarity on rights to assets and assignment of liabilities among owners (the government), managers, and creditors. This adjustment will require a clear separation between firms and the government, legally and operationally.
- Organizational structures that provide a reasonable balance of interests among owners, creditors, and managers.

Until passage of the State-Owned Enterprise Law in 1988, distinguishing the rights and obligations of the government (as owner) from those of the firm, firms belonged to the government and were even part of government departments; their losses were covered through budgetary allocations or directed bank loans. The government has begun separating some commercial activities from government bureaus, though much remains to be done. In particular, inefficiencies must no longer be financed explicitly or implicitly through the banking or budgetary systems.

As for organizational structure, the government has chosen to adopt the modern corporate form, in which the owner elects a board of directors to oversee the daily operation of the firm by professional managers.[15] The government has also set up an elaborate state asset management system to monitor the performance of state enterprises and ensure that their assets are preserved. But progress has tended to lag behind principles. The state asset management system is short on qualified staff, budget, and motivation. None of its institutions receives timely, accurate financial information. It is not clear whether, even with time, the new arrangements will ensure that enterprises are properly managed in an increasingly competitive environment.

The government has little time to improve the efficiency and competitiveness of the 1,000 large enterprises. Many will be exposed to international competition once China joins the World Trade Organization (see chapter 7). Still, it is best to make tough decisions now, while firms still have time to cut costs, reconfigure their operations, and improve competitiveness. Here more than anywhere else, the government should consider diversifying ownership so that firm-level changes are based on purely commercial criteria and that risks are spread across many shareholders. Since change is inevitable, delays will only add to costs, force defensive adjustments under market pressures, and contribute to the bad debts of banks.

If the government will be stretched in managing the priority 1,000 state enterprises, what of the remaining 304,000? It has stated its intention to loosen controls over them, implicitly granting them the freedom to organize leases, mergers, sales, and (if all else fails) liquidations. But it could go further, withdrawing completely from small and medium-size firms. It has already pulled back a long way: small and medium-size state enterprises contributed only 6 percent of industrial output in 1994, down from 36 percent in 1978. In many industries state enterprises contribute less than a fifth of gross output (figure 3.4).

Diversifying ownership would not require the government to sell its entire ownership stake at once. Initially, it could dilute its holdings and become a minority shareholder, leaving management to the new nonstate owners. For the government this would be a win-win solution as it shifts from being the sole owner of loss-making enterprises to being part owner of profitable ones. To maintain credibility, this move should be accompanied by a strict policy forbidding new state enterprises in contestable sectors (box 3.3).[16]

The government could also proceed faster. In two counties—Zhucheng in Shandong Province and Shunde in Guangdong Province—most state enterprises have already been transferred to the nonstate sector. Progress elsewhere has varied, with coastal provinces ahead of most others. Areas that lag will find that the costs of state ownership grow, eventually corroding the foundations and credibility of reforms.

Transforming the banking system

Intertwined with state enterprises is China's large financial system (box 3.4). Its core consists of four state commercial banks that together account for more than 90 percent of bank assets and two-thirds of financial assets. The condition of these banks affects the health of the entire economy. So it is cause for concern that their financial performance is weakening; their accounting,

FIGURE 3.4

Ubiquitous state enterprises

Percentage share of state-owned enterprises in gross output by industrial subsector

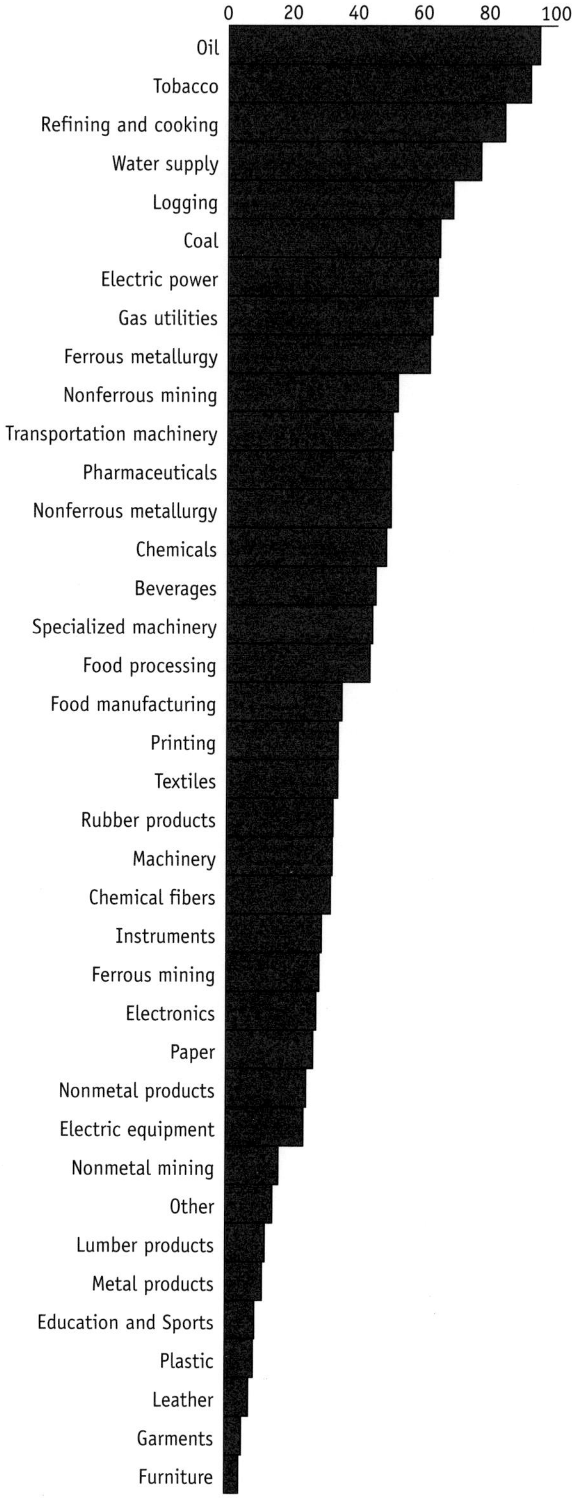

0 20 40 60 80 100

Oil
Tobacco
Refining and cooking
Water supply
Logging
Coal
Electric power
Gas utilities
Ferrous metallurgy
Nonferrous mining
Transportation machinery
Pharmaceuticals
Nonferrous metallurgy
Chemicals
Beverages
Specialized machinery
Food processing
Food manufacturing
Printing
Textiles
Rubber products
Machinery
Chemical fibers
Instruments
Ferrous mining
Electronics
Paper
Nonmetal products
Electric equipment
Nonmetal mining
Other
Lumber products
Metal products
Education and Sports
Plastic
Leather
Garments
Furniture

Source: China Statistical Yearbook 1996.

risk management, and credit analysis systems are woefully inadequate; and the quality of their portfolio is unknown. What is more, their capital-asset ratios are low and declining. If, as the government estimates, their nonperforming assets are equivalent to about 20 percent of their portfolios,[17] their net worth is actually negative.

Unfortunately, this state of affairs is unlikely to improve for two reasons. First, despite recent reforms, state banks are not yet completely free to lend according to commercial criteria. About a third of their investment lending is allocated to projects selected by the State Planning Commission, and the rest is subject to considerable informal government influence, especially in the provinces. As a result the creditworthiness of borrowers and the commercial viability of projects are often not important considerations in lending decisions.

Second, interest rates are controlled, and the spread between deposit and lending rates is limited by central bank decree. Consequently, even according to official statistics, the rate of return on banks' assets has been falling and turned negative for the first time in 1996. This deterioration occurred despite the fact that many banks treat accrued (but often unpaid) interest as paid income, make negligible provisions for bad debt (based on instructions from the Ministry of Finance), and raise earnings through various unofficial add-on fees and service charges. Over time, the result has been a steady decapitalization of the banks.

Transforming the state banks into genuine commercial banks is crucial if China is to avoid a major bank insolvency and if market forces are to shape the allocation of financial resources in China. Many of the necessary steps are incorporated in the government's long-term strategy (box 3.5). It calls for a phased transfer of government-directed lending to the policy banks, whose decisions will hew to government imperatives. Phasing will give state enterprises time to adjust to the new regime, but it should not extend beyond 2000.

But the most important policy change required is a reduction in government influence over the allocation of loans. This would permit more competition between banks, creating pressure to improve the quality of loans and increase the efficiency of intermediation. A nucleus of potential competition already exists in the five

BOX 3.3

If it is glorious to be rich, is it OK to be private?

Considerable government attention has been given to reforming state enterprises in China. But what of China's 1.5 million collectively owned enterprises? Once the powerhouse behind the economy's growth and employment (see chapter 1), their dynamism has waned (see figure). After generating some 17 million jobs in 1993, they created just 1.4 million in 1994 and 1995, while private and individually owned enterprises created 6.6 million new jobs.

There are several possible explanations for the decline. One is that the fuzzy ownership structure of collectives, appropriate for the early stages of reform, is no longer conducive to speedy growth. Many collectively owned firms are founded by groups of entrepreneurs in rural areas who invite the local government to share control rights. In return they gain access to key factors of production (land, capital) and services (transport, infrastructure) and government protection of property rights. This shared ownership tends to blur the division of control over the decisions and profits of the firm. Over time such vagueness in the allocation of property rights tends to erode efficiency and flexibility in operations. Increasingly, however, the growing acceptance of nonstate firms operating without any government participation has generated a sharp increase in private and individually owned firms. Although most of these firms are small and concentrated in services, a growing proportion is in manufacturing.

The government's pragmatic recognition of these developments is reflected in the Township Enterprise Law, which went into effect on January 1, 1997. Article 10 states that the "property rights of township enterprises set up by groups of farmers or individual farmers are owned by the investors." Perhaps the time is ripe for senior government leaders to announce that it is OK to be private. Who knows? Such an announcement could unleash the same sort of energy as was generated when the late Deng Xiao Ping said that to be rich is glorious.

The growing share of employment in "other" enterprises, 1983–95

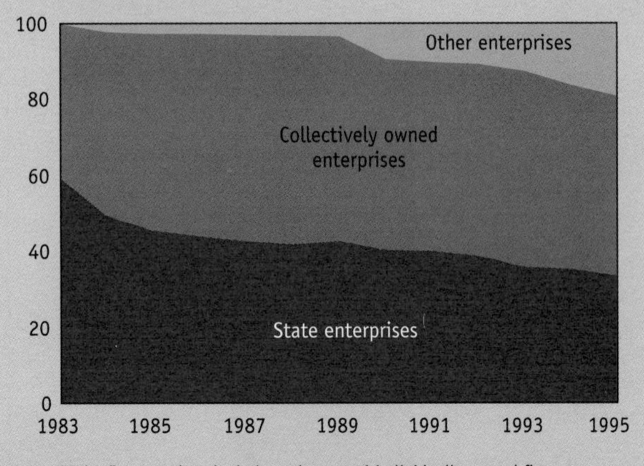

Note: "Other" enterprises includes private and individually owned firms as well as joint ventures and foreign-funded firms.
Source: China Statistical Yearbook, various issues.

nationwide commercial banks (including the recently established Hua Xia Bank and Ming Shen Bank) and eight regionally based commercial banks. But the government should consider expanding the field of competitors. More new entrants could be licensed, provided they satisfy the central bank's strict requirements. The government could also consider splitting the largest state commercial banks into several entities, perhaps initially along regional lines. Finally, the government could, over the longer term, gradually permit the entry of foreign banks, but only after the domestic banking system is competitive and working well under its new framework of laws and regulations.

Reform of interest rate policy will be an essential but treacherous feature of China's financial development. The government has already taken steps in the right direction by reducing the number of officially determined interest rates, introducing treasury bill auctions, adjusting interest rates more often in line with inflation, and initiating open market operations.

BOX 3.4

China's financial system . . .

A variety of institutions populate the financial sector. At the apex is the People's Bank of China, established in 1984 as a separate central bank responsible for making monetary policy and supervising the financial system. There are four huge state commercial banks, created initially as specialized banks, and three policy banks. In their shadow are fourteen smaller commercial banks, most of them established in the last six years (two in the past year).

There is also a growing fringe of nonbank financial institutions, including thousands of rural and urban credit cooperatives operating as near-banks, more than 300 trust and investment companies, and an expanding presence of foreign financial institutions in the form of representative and branch offices. Completing the financial system are two securities exchanges at Shanghai and Shenzhen, twenty-three regional securities trading centers, two electronic securities trading networks (STAQS and NETS), and myriad securities companies, brokers, dealers, and underwriters.

Once the government is satisfied with progress in commercializing state banks, the next logical step is to allow greater flexibility in setting lending (but not deposit) rates. The central bank could allow commercial banks to set lending rates freely on a small portion of their portfolios, which they could expand with time, or it could gradually widen the bands within which lending rates could fluctuate.

How the banking system evolves over the next twenty-five years will hinge on the success of these reforms. The rewards for success are high. A recent study shows that well-functioning financial systems tend to stimulate and sustain growth because they help pool savings and allocate them to their most productive use, subject to appropriate safeguards.[18] But the government is also aware that the risks of failure can be high and the costs punishing. A World Bank study has identified more than 100 major episodes of bank insolvency in ninety developing and transition economies from the late 1970s to 1996. In twenty-three of the thirty countries for which data were available, the direct losses sustained by governments exceeded 3 percent of GDP and in one case reached more than 50 percent (figure 3.5).

The financial sector has been the soft underbelly of China's reforms. It is the sector most vulnerable to shocks. The authorities must be as relentless in their pursuit of reforms in this area of the economy as they have been in others.

Deepening capital markets

China's capital markets are subject to the same constraints and controls as other parts of the financial sector. The credit plan sets quotas on how much equity and securitized debt can be issued in a given year. The primary market for equities and corporate bonds is distributed by region, limiting capital market access in some regions more than in others.

Thus the relatively small role of capital markets (figure 3.6) is not a product of market forces but of administrative decree. The authorities fear that unrestrained capital market development would drain resources from the state commercial banks. So for years to come, the growth of China's capital markets will remain

FIGURE 3.5

The direct costs of banking crises can be high

Percentage of GDP

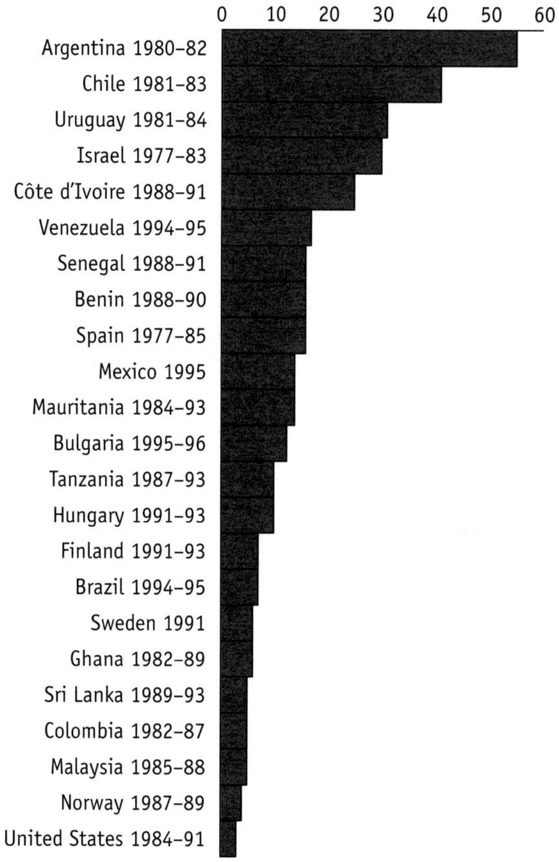

0 10 20 30 40 50 60

Argentina 1980–82
Chile 1981–83
Uruguay 1981–84
Israel 1977–83
Côte d'Ivoire 1988–91
Venezuela 1994–95
Senegal 1988–91
Benin 1988–90
Spain 1977–85
Mexico 1995
Mauritania 1984–93
Bulgaria 1995–96
Tanzania 1987–93
Hungary 1991–93
Finland 1991–93
Brazil 1994–95
Sweden 1991
Ghana 1982–89
Sri Lanka 1989–93
Colombia 1982–87
Malaysia 1985–88
Norway 1987–89
United States 1984–91

Source: Caprio and Klingebiel 1996; World Bank 1997d.

BOX 3.5

. . . and its recent reform

In 1993 the government announced wide-ranging reforms in the financial sector. Since then it has:

- Created three policy banks to handle lending for government-directed investments and initiated a program to transform China's four enormous state banks into commercial banks.
- Passed laws clarifying the authority of the central bank and the rights and obligations of commercial banks.
- Granted state commercial banks increasing autonomy in lending decisions and encouraged the entry of new commercial banks.
- Introduced new, indirect techniques of monetary management, including a more active interest rate policy, a unified interbank market, auctions for treasury bill issues, asset-liability ratios for commercial banks, and open market operations.
- Severed ownership links between banks and nonbank financial institutions and tightened regulations governing China's two stock exchanges.
- Initiated a modern payments system.

FIGURE 3.6

China's relatively small capital market, 1995
(Stock of financial savings)

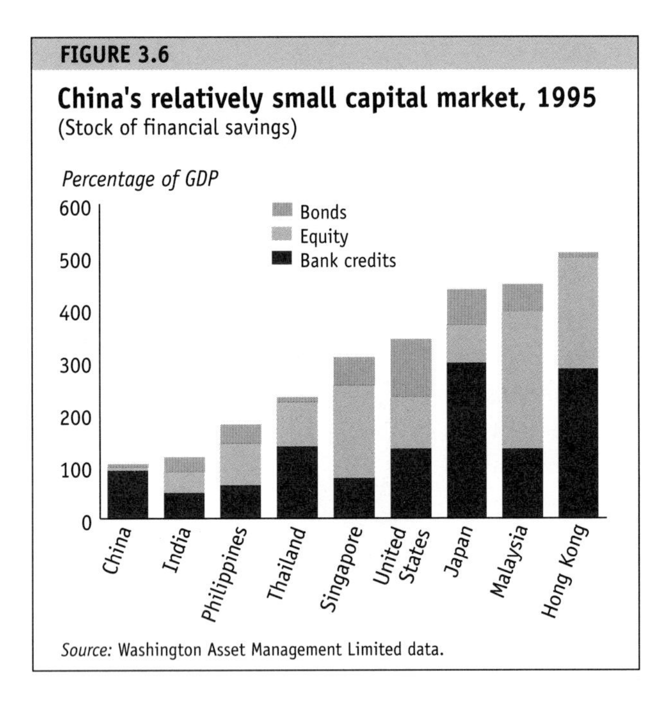

Source: Washington Asset Management Limited data.

FIGURE 3.7

Frenzied trading at Shanghai and Shenzhen, 1994

Ratio of trading volume to market capitalization

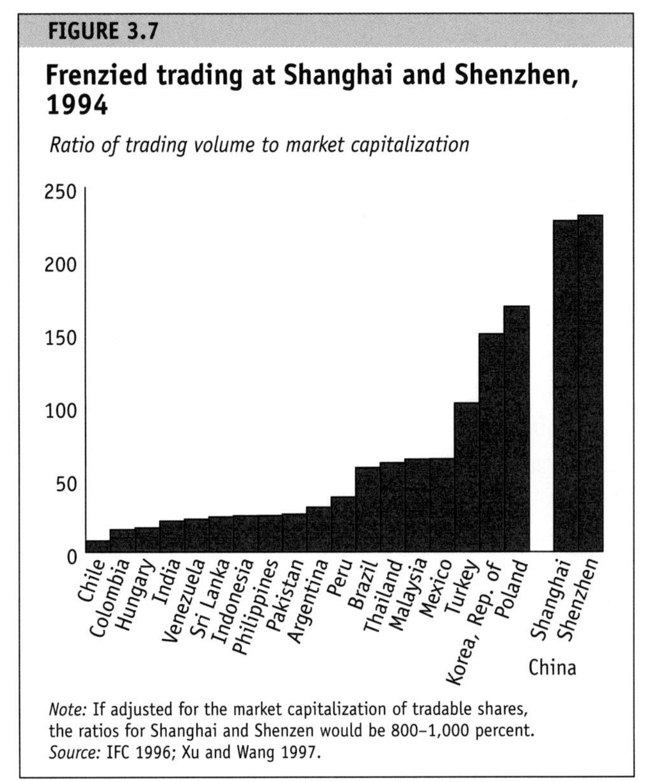

Note: If adjusted for the market capitalization of tradable shares, the ratios for Shanghai and Shenzen would be 800–1,000 percent.
Source: IFC 1996; Xu and Wang 1997.

hostage to progress in the development of a sound and competitive banking system.

Administrative constraints tend to affect primary markets more than secondary markets. Even so, secondary markets have problems of their own. Share prices are very volatile in the Shanghai and Shenzhen stock exchanges, and trading volumes are unusually high relative to the size of market capitalization (figure 3.7). This is because both markets are young, with an underdeveloped legal and institutional infrastructure. Like other emerging markets, they suffer from low liquidity, limited disclosure, bunching of public offerings, insider trading, and a circumscribed role for competitive underwriters and primary dealers.

Compounding these shortcomings is the limited participation by wholesale and institutional purchasers of securities. China's contractual savings are equivalent to about 3 percent of GDP, well below the 18 percent in the Republic of Korea, 48 percent in Malaysia, and 78 percent in Singapore. The insurance industry has only a short history in China, and pension funds are small because of the preponderance of pay-as-you-go schemes.

Finally, government oversight of the capital market is weak. Although securities laws and regulations are guided by international principles, several crucial aspects of capital markets still require codified rules (table 3.4). To make matters worse, regulatory responsibility for securities markets is splintered functionally, institutionally, and regionally and is divided between municipal and central authorities. As a result some regulations—for corporate bond trading, securities dealers, and institutional participants—are not overseen by any government institution.

Resolving these problems quickly is important, and there is no time to lose. Even if capital markets grow at more than twice the rate of projected GDP growth (in real terms), by 2020 the value of stocks and bonds relative to China's economy would only approach that in India's capital market today (74 percent of GDP).

Capital markets could play a major role in China's financial and economic development:

• Through capital markets, investors can monitor and control the corporate users of their capital. Such oversight is an essential element in state enterprise reform because it helps dilute government ownership and transfers a portion of the risks from banks to the new owners. The example of "red chip" companies in Hong Kong also shows how listings on other stock exchanges hold state enterprise managers in China to increased scrutiny, higher management standards, and greater transparency.[19]

TABLE 3.4
Gaps in China's securities legislation

Country	Securities law	Established securities and exchange commission	Disclosure regulation	Listing requirements	Insider trading regulation	Compensation fund	Takeover regulation
Argentina	●	●	●	●	●	●	●
Brazil	●	●	●	●	○	●	●
Chile	●	●	●	●	●	●	●
Korea, Rep. of	●	●	●	●	●	○	○
Malaysia	●	●	●	●	●	●	●
Mexico	●	●	●	●	●	●	●
Thailand	●	●	●	●	●	●	●
China	○[a]	●	○	○	○	○	○

a. Elements of a securities law are covered in other laws and regulations.
Source: World Bank 1997c.

• Flourishing capital markets would be a significant intermediary for long-term investment, especially for infrastructure. Chapter 4 describes how the introduction of a new pension system would mobilize massive amounts of investable funds. How efficiently (and safely) they are invested will depend largely on the depth, stability, and efficiency of the capital market.

• Thriving capital markets would provide savers with new savings instruments, helping them to diversify risk and boost returns. Such instruments would be an important piece in the mosaic of policies designed to encourage the savings that are so essential for sustainable growth (see chapter 2).

• Well-functioning capital markets would improve the pricing of financial assets, contributing to greater efficiency in the allocation of resources. In this regard China's capital markets could be far more valuable than their relative size would indicate.

Bonds are the least developed part of China's capital markets (figure 3.8). Developing the bond market will be crucial for tapping China's massive domestic savings and matching savings to its equally large infrastructure needs. Other developing countries—Chile, Malaysia, Thailand—have shown how these goals can be achieved. First and foremost, bond issuers must meet minimum standards of capital structure, financial strength, corporate governance, and management quality. Second, these companies should be rated by reputable credit rating agencies. Third, institutional investors should be encouraged to buy bonds, possibly by allowing them to invest in blue-chip corporate bonds up to specified limits and certainly by making corporate

FIGURE 3.8
Bonds—the least developed segment of China's capital markets

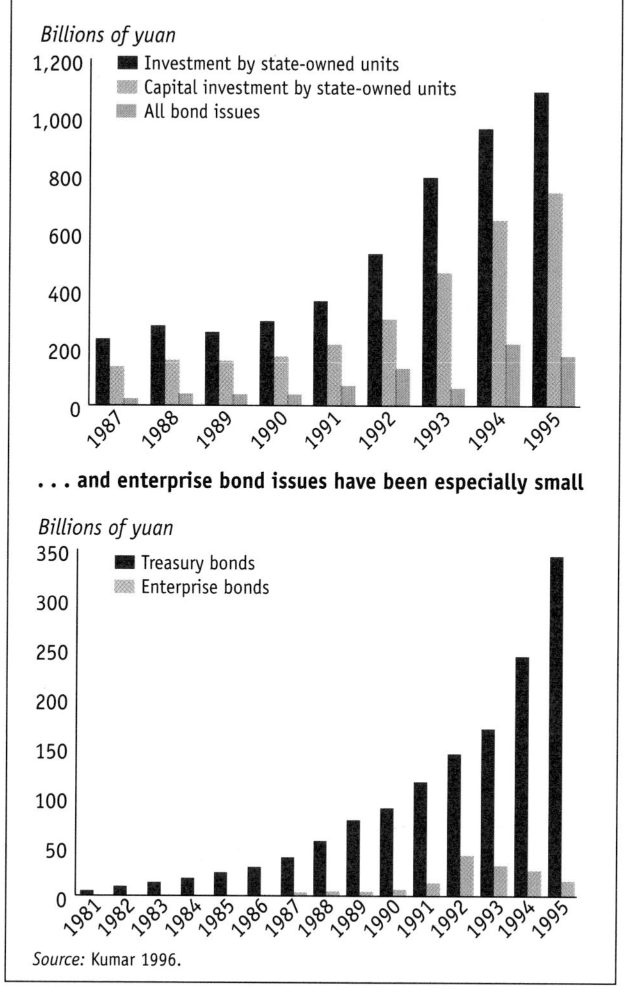

Source: Kumar 1996.

bonds fully tradable. Fourth, bond issues could be auctioned, so that financial institutions can become wholesale buyers.

China is on the threshold of exciting developments in infrastructure financing. With the help of multilateral institutions, including the World Bank Group, it is developing the legal and competitive framework required to issue asset-backed securities. Doing so will allow an increasing portion of its infrastructure needs to be financed from limited recourse private capital, domestic and foreign. The potential is enormous. A recent World Bank study estimates future demand for infrastructure in China to be as much as $75 billion a year over the next decade—well above the stock of outstanding enterprise bonds today of about $2 billion.[20]

To accelerate the growth of project financing, the government should also consider using credit enhancements, targeted risk and policy guarantees, partial subsidies, and (occasionally) direct participation. Another innovation that could be tried is infrastructure development funds that purchase securities from a pool of infrastructure projects. Pakistan's Private Sector Energy Development Fund and Jamaica's Private Sector Energy Fund are examples.

Housing finance also promises to grow rapidly in China in the next century. In high-income countries mortgages are one of the largest segments of capital markets. China still has little or no mortgage lending. Once it has a high-quality primary market for mortgages, secured though legally backed collateral and based on market interest rates, this market could be supplemented by a thriving secondary market.

Such a development would bring many benefits. International experience suggests that specialized housing finance institutions have weaknesses of their own.[21] The risks associated with liquidity and concentrated portfolios are best dealt with through an efficient secondary market. Increasingly, countries are creating secondary mortgage markets to allow banks to offload their mortgages and spread their risks (Australia, Malaysia, and Thailand are examples in the Asia-Pacific region). Furthermore, the securitization of mortgages helps deepen and develop capital markets by matching long-term savings with long-term borrowing. In many ways the secondary mortgage system could mirror developments in the pension and insurance systems, as both come together to deepen the debt market.

Making markets work

Reforming state enterprises and state commercial banks involves a gradual government withdrawal from direct interventions in product and financial markets. But in no sense should this withdrawal be construed as an abdication of responsibility for markets, because the government has a new obligation to provide the institutional and policy frameworks that help markets work. Partnership and complementarity with markets need to supplant command and control.

Four areas are of particular importance. First, establishing a rule of law that is supportive of market institutions and protective of property rights. Second, developing the capacity to ensure macroeconomic and financial stability, even as potentially destabilizing reforms are implemented. Third, promoting competition in product markets to encourage innovation, efficient resource allocation, and financial discipline. And fourth, harnessing new technology in ways that work with markets, not against them.

Establishing the rule of law

As the government extricates itself from markets, legal norms and procedures must substitute for direct government control over economic decisions. Simply put, economic reforms have made legal rules matter.

China has made considerable progress in developing legal norms that correspond to the needs of a market economy. Indeed, the process has accelerated in recent years, with as many laws (152) scheduled for consideration in the 1993–98 session of the National People's Congress as were enacted between 1949 and 1992. At the same time, there has been a rise in the use of the judicial system to resolve disputes, an emergence of a legal profession, and a growing respect for the independence of the judiciary.

Yet much remains to be done. It may take decades for the rule of law to establish a firm foothold in China. Doing so will require a legal infrastructure that can impartially implement the evolving legal framework, control corruption, and foster a legislative system based on public participation and consent.

Ten years ago China's economic laws were few and the gaps many. Today the maturing of the legal system is reflected by laws so numerous they defy easy summary. Yet the system operates poorly. The reason is not a lack of laws. It is inadequate enforcement. Enforcing the law is the biggest challenge that China's legal system will face in the next century. In key areas of recent legislation the law of the land is quite different from the law on the ground.

One remedy is to develop the legal profession. The new Lawyers Law, which went into effect in January 1997, could give a strong impetus to growth of the profession. The law aims to promote a better national qualification and licensing system, continuing legal education, a self-regulating bar, and rules of ethics. The most radical change is that lawyers are no longer state legal workers in advisory offices. They can now organize as state-funded, cooperative, or partnership law firms, law offices, or sole practitioners. A recent count identified 7,200 firms nationwide, most of them organized as partnerships.

The next big step will be to improve legal education, which was revived in 1978 after more than a decade of inactivity. Today numerous full-time law faculties and other institutions provide legal education and training. But a large portion of China's tens of thousands of lawyers have no formal legal training. Thus training needs include more extensive legal education, modernized curriculums, and practical and ethical training in the actual implementation of laws.

Corruption—the use of public office for private gain—is a growing problem in China. In light of international evidence that corruption may be as damaging to growth as political instability,[22] the government's concern about corruption is entirely appropriate.[23] In 1996 alone 65,424 people involved in embezzlement and bribes were prosecuted, including many senior officials in the government and the Party.

More could be done. The war against corruption will need to be waged on several fronts. One is through criminal prosecution, to increase the risk of detection and punishment. But China will also have to eliminate or reduce incentives for corrupt practices, including:

• *Reducing the discretionary power of government officials by increasing market competition.* International experience suggests that countries with heavy government intervention and little competition tend to be more corrupt.[24] Where possible, the government should use market-friendly, competitive approaches to provide public services. Nonstate companies could compete with government in delivering public goods and services. Such approaches give citizens a choice and curb the discretionary power of public officials.

• *Increasing transparency in government finances.* The more transparent and streamlined are government procedures, the fewer the opportunities for corruption. Transparency helps keep government officials accountable for their actions. For example, making unaudited, unsupervised extrabudgetary funds available to officials invites misappropriation. The danger can be minimized by better oversight, better information on targets and achievements, and better channels for listening to citizen groups.

• *Building an honest and competent civil service.* Choosing new recruits carefully based on academic and professional qualifications will pay handsome dividends, as will reducing salary differentials between the civil service and the rest of the economy (box 3.6). It might also be useful to establish an independent statutory institution, such as Hong Kong's Independent Commission Against Corruption, to act as watchdog.

BOX 3.6

Nourishing honesty

In 1723 Emperor Yongzheng faced widespread corruption in his tax administration. As part of sweeping fiscal reforms, he imposed a surcharge on basic land taxes and used the revenue to finance large salary increases for revenue officials. These bonuses were known as yang lian—money to nourish honesty.

Ensuring stability

To flourish, markets need a background of macroeconomic stability. Since 1978 the government has usually relied on direct, administrative measures to cool the economy when it threatened to overheat. Similarly, the stability of financial institutions was ensured through direct support by the central bank. In the future, however, the government faces the more difficult task of maintaining macroeconomic and financial stability using indirect, market-friendly measures. Success will depend on the extent to which banks and enterprises have become responsive to market signals.

The central bank's ability to control monetary aggregates has been enhanced by the Central Bank Law, which forbids lending and overdrafts to the government. In addition, the People's Bank of China has occasionally recalled loans to financial institutions, refined its reserve requirements to suit individual banks' needs, opened a rediscount window to refinance strategic projects, and called for special deposits. These developments allowed the bank to begin limited open market operations in April 1996.

Improving the central bank's monetary management is central to future economic stability. The erosion of the credit plan as an effective instrument of monetary control means that the central bank will need other ways to control monetary aggregates. It already uses reserve ratios, asset-liability ratios, and the rediscount facility to affect the liquidity of the financial system. It will also need to make more extensive use of open market operations. In doing so, it will have to guard against sudden and excessive withdrawal of liquidity from the financial system. To increase the availability of securitized assets for sale, the central bank could consider securitizing its large stock of loans to the government and selling these on the market. Doing so may have budgetary consequences if government interest payments have to rise.

The introduction of open market operations and unification of the interbank market are already producing greater flexibility in short-term interest rates. Controlling aggregate demand more effectively will require these interest rate changes to ripple through the maturity spectrum of bank assets and liabilities. In addition, enterprises must be motivated to keep costs down—which means they must not receive subsidies if they operate at a loss.

As the financial system changes, there is always the danger that risky activity by banks and nonbank financial institutions will generate systemic shocks. The collapse of the Shanghai International Securities Corporation and, more recently, the China Agricultural Development Trust and Investment Corporation were warning signs. To improve the health of banks and nonbank financial institutions, their managers and their official supervisors must pay more attention to prudential norms and less to traditional compliance with key targets and ratios. This shift in emphasis will require modern accounting systems and new financial reporting methods, a uniform risk-based loan portfolio classification system, and a focus on the overall risk borne by individual financial institutions. Some progress has already been made on these fronts, but there is still a long way to go.

Good supervision will provide the added benefit of ensuring the safety and soundness of the financial system. Training for supervisors at all levels would make a valuable contribution. The frequency of onsite inspections will also need to be increased at the provincial, city, and even county levels. All banks will have to implement accounting and auditing standards.

Making markets more competitive

Perhaps the most important reason the Chinese economy performs so well is that many goods markets are quite competitive. Some observers have also argued that "market-preserving federalism" has encouraged competition between provinces by motivating local governments to foster local prosperity.[25]

Still, more could be done to intensify competition. Provincial governments could phase out remaining barriers to interprovincial trade—especially in agriculture, where state enterprises control the bulk of distribution and sales (see chapter 5). Similarly, provincial governments could remove any restrictions requiring local state enterprises to buy from other local enterprises. The gains from extra interprovincial trade could be comparable to the benefits of additional international trade.

The benefits of competition are readily apparent in sectors with low barriers to entry and exit. Their growth rates have been uniformly higher than in industries dominated by state firms. The government could take steps to encourage the entry of nonstate companies, domestic and foreign, into manufacturing and infrastructure. Phasing out the tax advantages that foreign investors enjoy, especially in special economic zones near the coast, would level the playing field for domestic and foreign investors and encourage investments in the interior, western provinces. Similarly, eliminating the setting of interest rates by class of ownership would remove one of the advantages enjoyed by state enterprises at the expense of nonstate firms.

China's current industrial policy emphasizes support for "pillar" industries (box 3.7). However, it is often difficult to say what this policy really means. The government uses many instruments to influence the pattern of industrial production: controlled interest rates, pro-

BOX 3.7

China's "pillar" industries

A new industrial policy announced on March 25, 1994, emphasized the development of "pillar" industries. The Chinese arrived at the definition of pillar industries by analyzing the stages of industrialization along the lines identified by Chenery and Syrquin (1975). Thus light industries constitute the first stage of industrialization, basic heavy industries the second, and pillar industries the third.

The government has designated five pillar industries: machinery, electronics, petrochemicals, automobiles, and construction. These industries were chosen because they are expected to face a high income elasticity of demand, enjoy substantial economies of scale, result in significant backward and forward production linkages, possess potential for high productivity growth, and reflect China's comparative advantage. The hope is that they will eventually account for 5 percent of GDP (or 8 percent of industrial output), increase their share in international markets, reach international quality standards quickly, and become profitable.

tection against import competition, tax and price policies, rules on procurement, and even unofficial levies. The combined effect of these measures is difficult to assess, but it is quite likely that they work at cross-purposes and in unintended ways that distort development.

Government officials recognize that China may not be able to follow the role models of Japan and the Republic of Korea for industrial policy. For one, entry into the World Trade Organization will constrain their freedom to direct credit or budgetary subsidies to specific industries, and protection from import competition can only be very limited. Trading partners today are considerably less tolerant of such practices than they were in the 1950s and 1960s. For another, China possesses neither the capabilities nor the conditions that Japan and Korea enjoyed. China's government structure, for example, is complex and compartmentalized, discouraging communication across ministries and complicating coordination across tiers of government. Administrative decentralization has rekindled an earlier bias toward provincial self-sufficiency, so provincial goals often conflict with national goals. Procedures and instruments of indirect management have not yet fully replaced those of direct management. The fiscal position is fragile, and the banking system is delicate. The tender shoots of a genuine private sector, although sprouting in increasing numbers, have yet to take hold and mature.

This is not a setting in which activist, wide-ranging industrial policies are likely to be successful. Indeed, in a fast-changing international environment such policies could do more harm than good. China has discovered that when the ambition of industrial policies outstrips government capability, the results can be costly and difficult to undo. In the one area in which it has aggressively pursued an industrial policy—automobiles—China has found itself burdened with significant excess capacity and large numbers of small, inefficient plants. Having been burnt, it is reassessing its approach toward industrial development. The focus appears to be shifting toward policies with a lighter touch—especially policies that are inexpensive and that support nonstate development. These include regulatory policies that create a level playing field for all firms and the provision of intraindustry public goods such as information, infrastructure, quality standards, and industry associations. These actions, if implemented, would be steps in the right direction, and their benefits should not be underestimated.

Harnessing new technology

Apart from measures to intensify competition, industrial policy can help develop and harness technology. In the long term rapid growth will depend heavily on China's ability to acquire, adapt, and master new technology.

China's Ninth Five-Year Plan and Fifteen-Year Perspective Plan emphasize the importance of intensive growth, which involves increasing productivity, lowering costs, and introducing higher-quality products. Chinese planners recognize the power of technology in sustaining growth over the long term and the importance of government's role in ensuring that the promise of technology

TABLE 3.5

Lawsuits on intellectual property rights in China, 1991–95

Legal issue	Lawsuits	Prosecutions
Patent	3,083	2,737
Copyright	2,600	2,429
Trademark	907	789
Technology transfer	6,812	6,805
Commercial know-how	2,141	2,100
Criminal offenses[a]	1,690	1,676
Foreign related[b]	192	170

a. Cases where criminal action or violence was a factor.
b. Cases where one of the parties was a foreign entity.
Source: Supreme Court of China, Intellectual Property Rights Office.

is fulfilled by developing a market for ideas and supporting it when it fails to operate effectively.

Firms and individuals will make less of a technological effort than is desirable if they are unable to reap the benefits. Thus protecting intellectual property rights is of prime importance. China has enacted and revised laws on patents, trademarks, and copyright, and has done much to enforce them. The number of lawsuits and prosecutions under these laws has grown substantially (table 3.5), and there are now specialized intellectual property courts throughout the country.

The government has also acted to create a market for ideas, "plunging the scientific community into the sea of markets."[26] Before reforms, the administrative apparatus transmitted findings from research units free of charge. But technology has now been redefined as a "commodity" in China. To establish markets for this commodity, the government has encouraged technology trade fairs at which producers of research and development can sell their wares to potential buyers. In 1995 the country held 939 such fairs at which 14,686 contracts were signed for a value of 5.8 billion yuan.

Of course, the market for ideas does not stop at China's borders. In many cases it will be more economical to acquire new commercial technology from abroad than to develop it at home. Much of this technology is embodied in capital equipment, such as turnkey projects or imported capital goods. In other instances it is packaged along with equipment finance and management, as in foreign direct investment. And in others technology comes "unbundled," through technical assistance or technology licenses.

The government should ensure that the costs of acquiring foreign technology are as low as possible and that the technology is disseminated domestically as efficiently as possible. Lower barriers to importing capital goods would help by encouraging equipment investment and technology imports, which in other countries tend to be associated with faster growth.[27] More transparent and streamlined foreign investment procedures would also encourage new foreign investors. And more liberal rules on technology licensing agreements would encourage closer links between technology suppliers abroad and producers at home.

In many cases, though, the market for ideas functions imperfectly. The owners of ideas may not be compensated for the goods they provide, so are reluctant to make their ideas available. Some ideas have long gestation periods, and their benefits are not readily apparent for many years. Government intervention may be needed to ensure that there is an appropriate supply of new ideas and technologies.

To remedy such market deficiencies, China's government has focused on basic research. Its "climbing" program for fundamental research is designed to keep pace with world advances in key areas of information technology, life sciences, and genetic engineering. Policymakers have invited high-technology firms to locate in more than 100 science and technology parks that provide infrastructure, priority access to finance and imports, and tax privileges. Today more than 13,000 enterprises are located in such parks, employing over 1 million people. In addition, in 1995 the government established sixty-nine national engineering centers as a link in commercializing research findings. These centers are funded by the public sector, but the intention is eventually to make them commercially independent.

Notes

1. Strictly speaking, the reference here is only to in-budget industrial enterprises.

2. Comparative data are from Levin (1991), who used a sample of eighteen industrial countries and twenty-two developing countries for which data on general government expenditures are available in the International Monetary Fund's (IMF) *International Finance Statistics*. Data are averages over three years ending in 1987 or 1988.

3. As mentioned later in this chapter, the three quasi-tax revenues were incorporated into the budget in 1997.

4. World Bank (1996b).

5. The relative price of services tends to rise with higher real per capita incomes.

6. This estimate was provided by the State Tax Administration and was based on the 1992 input-output table.

7. Extrabudgetary expenditures are recorded in data published by the State Statistical Bureau. But other government expenditures escape recording and quantification altogether, at both the national and provincial levels.

8. The consolidated government deficit is defined as the sum of the budget deficit and the volume of lending by the central bank specifically to finance government-directed investments. The authorities have indicated that this deficit can range from 60 to 80 percent of total lending by the central bank to the banking system.

9. Hofman (forthcoming).

10. Some extrabudgetary expenditures—such as hospital expenditures financed from cost recovery—need not be incorporated into the budget.

11. These figures refer to all state enterprises, not just industrial state enterprises.

12. World Bank (1996a). Only part of this increase in losses can be explained by the adoption of a new accounting system.

13. Jiang (1995).

14. Qian and Weingast (1997)

15. World Bank (1997a).

16. As noted earlier, a contestable sector is one in which there is relatively costless entry and exit. Although most Chinese industries are contestable, the government initially may want to focus on such industries as specialized machinery, printing, rubber products, machinery, chemical fibers, instruments, electronics, paper, nonmetal products, electric equipment, lumber products, metal products, educational and sports products, plastic products, leather, garments, and furniture.

17. According to People's Bank of China Governor Dai Xianglong, 8 percent of outstanding loans at state banks are more than three years overdue and another 12 percent are less than three years overdue (Faison 1996).

18. King and Levine (1993).

19. "Red chip" companies are mainland Chinese state enterprises registered in Hong Kong and listed on the Hong Kong stock exchange.

20. World Bank (1996c).

21. World Bank (1997d).

22. Mauro (1995).

23. Li (1996).

24. World Bank (1997e).

25. Qian and Weingast (1997).

26. China State Science and Technology Commission (1996).

27. Mody and Tilmaz (1991).

References

Blanc, J., and C. Brule. 1993. *Les nationalisations françaises en 1982*. Paris: La Documentation Française.

Caprio G. Jr., and D. Klingebiel. 1996. "Bank Insolvency: Bad Luck, Bad Policy, or Bad Banking?" In M. Bruno and B. Pleskovic, eds., *Annual World Bank Conference on Development Economics 1996*. Washington, D.C.: World Bank.

Chenery, H., and M. Syrquin. 1975. *Patterns of Development, 1950–70*. New York: Oxford University Press.

China State Science and Technology Commission. 1996. "Technology Policy in China." Background paper prepared for the World Bank. Washington, D.C.

De Long, J. B., and L. H. Summers. 1991. "Equipment Investment and Economic Growth." *Quarterly Journal of Economics* 106 (2): 445–502.

Faison, S. 1996. "Inflation Curbed but not Growth, China Asserts." *New York Times*, July 16.

Hofman, B. Forthcoming. "Fiscal Decline and Quasi-Fiscal Response: China's Fiscal Policy and Fiscal System, 1978–94." In OECD/CEPII/CEPR, *Different Approaches to Market Reforms: A Comparison Between China and the CEECs*. Paris.

IFC (International Finance Corporation). 1996. *Emerging Stock Markets Factbook*. Washington, D.C.

Jiang, Q. 1995. "State Asset Management Reform: Clarified Property Rights and Responsibilities." In H. Broadman, ed., *Policy Options for Reform of State-Owned Enterprises*. Washington, D.C.: World Bank.

King, R.G., and R. Levine. 1993. "Finance, Entrepreneurship, and Growth." *Journal of Monetary Economics* 32:513–42.

Kumar, A. 1996. "China's Domestic Bond Market and Infrastructure Investment." Paper presented at the seminar on Mobilizing Domestic Resources for Infrastructure Financing, Government of China and the World Bank, November 12, Beijing.

Levin, J. 1991. "Measuring the Role of Subnational Governments." IMF Working Paper 91-8. International Monetary Fund, Washington, D.C.

Li, Peng. 1996. "Report on the Outline of the Ninth Five-Year Plan for National Economic and Social Development and the Long-Range Objectives to the Year 2010." Speech delivered at the Fourth Session of the Eighth National People's Congress, March 5, Beijing.

Mauro, P. 1995. "Corruption and Growth." *Quarterly Journal of Economics* (August):681–712.

Mody, A., and K. Tilmaz. 1991. "Cost Reduction through Imports of Capital Goods." World Bank, Industry and Energy Department, Washington, D.C.

Qian, Y. 1995. "Enterprise Reform in China: Agency Problems and Political Control." Stanford University, Department of Economics, Stanford, Calif.

Qian, Y., and B. Weingast. 1997. "Institutions, State Activism, and Economic Development: A Comparison of State-Owned vs. Township-Village Enterprises in China." In M. Aoki, H-K Kim, and M. Okuno-Fujiwara, eds., *The Role of Government in East Asian Economic Development: Comparative Institutional Analysis*. New York: Oxford University Press.

World Bank. 1996a. "China: Reform of State-Owned Enterprises." Report 14924-CHA. China and Mongolia Department, Washington, D.C.

———. 1996b. *The Chinese Economy: Fighting Inflation, Deepening Reforms*. A World Bank Country Study. Washington, D.C.

———. 1996c. *Infrastructure Development in East Asia and Pacific: Toward a New Public-Private Partnership*. Washington, D.C.

———. 1997a. *China's Management of Enterprise Assets: The State as Shareholder*. A World Bank Country Study. Washington, D.C.

———. 1997b. *Financing Health Care: Issues and Options for China*. Washington, D.C.

———. 1997c. *The Road to Financial Integration: Private Capital Flows to Developing Countries*. A World Bank Policy Research Report. New York: Oxford University Press.

———. 1997d. "Strategic Framework for the World Bank Group for the Financial Sector." Presented to the World Bank Executive Directors by the Financial Sector Department, April 21, Washington, D.C.

———. 1997e. *World Development Report 1997: The State in a Changing World*. New York: Oxford University Press.

Xu, X., and Y. Wang. 1997. "Ownership Structure, Corporate Governance, and Corporate Performance: The Case of Chinese Stock Companies." World Bank Policy Research Working Paper 1794. World Bank, Economic Development Institute, Washington, D.C.

Zhang, T., and H. Zou. 1996. "Determinants of Provincial Income Growth in China." World Bank, Washington, D.C.

chapter four

Shaping a Competitive but Caring Society

Shaping a competitive economy requires market forces. Shaping a caring society requires government leadership. Striking the right balance between competition and compassion depends largely on a country's circumstances, traditions, and culture. In China economic reforms have led to impressive growth in incomes. But they have also led to steep increases in social and income inequality, with limited improvements in living standards for the poor. Looking ahead, China will need market forces to ensure that resources and people combine in ways that support continued and sustainable improvements in income and welfare. It will also need an active government to help its people manage the increased risks and uncertainties generated by the market, and to build support mechanisms for the most vulnerable.

The next twenty-five years will bring tumultuous change to China. If growth runs at the government's projected average of 6–7 percent a year, the economy will expand sevenfold. Thus China would experience in twenty-five years an economic transformation similar to that in Latin America over the past sixty-five years and in OECD countries over the past eighty.

Such big changes, compressed into so short a time, will stretch China's social fabric to the limit. Hundreds of millions of people will move in search of better jobs from agricultural to nonagricultural occupations and from rural to urban areas. Urban labor markets will also have to handle huge structural change within industries, especially shifts between state and nonstate firms. By 2020 a typical worker in China will not be a self-employed farmer but a wage employee on a terminable contract in either industry or services. Flexible rural labor markets and efficient rural-urban migration will be essential, particularly since growth of the labor force will be slowing considerably.

Greater mobility between jobs, sectors, and occupations will also require greater flexibility in education. Upgrading skills by increasing investments in general education and on-the-job training will ease structural change. And greater access to education in rural areas (especially for girls) will be central to reducing poverty and bringing the untapped potential of the poorest people into the economic mainstream.

The speed and size of these employment changes will foster both hope and uncertainty. Hope—of better jobs and rising incomes—is already real and palpable. But unemployment, migration, old age, and illness will present new challenges for China's people. Those who were most protected in the past—mainly urban workers in state-owned units—will find it especially difficult to adjust to the new reality.

As they grow richer, all societies demand formal mechanisms to manage such risks. In China this demand will grow very quickly, and will probably center on five issues where government action is needed:

- *The living standards of the absolute poor.* There has been a remarkable reduction in the number of poor people since reforms started. But eliminating poverty is becoming increasingly difficult. The absolute poor are usually entire communities that live in isolated, upland regions of interior provinces with few if any natural resources. Bringing them into the mainstream of economic development will require new approaches and better targeting.
- *Financial security for the elderly.* Because state enterprises are finding it increasingly difficult to meet their pension obligations, families are doing more to support their elderly members. But these informal arrangements are becoming a financial strain. The one-child policy means that a working couple usually has to support four retired parents, one child, and sometimes grandparents—an extraordinary burden, especially when people have to move in search of jobs. Without more formal arrangements, China may not be able to maintain its traditional reverence for the old.
- *Access to affordable health care.* Public health facilities are being squeezed by budgetary pressures, and the cost of health care is rising rapidly. Almost all health insurance is for urban residents, especially government and state enterprise employees. Rural areas are neglected, with the old, the young, and the poor suffering the most.
- *Biases against women.* By international standards, the status of women in China's labor market is relatively favorable. But international experience also shows that market reforms tend to hurt women disproportionately. This may be occurring in China as well.
- *High and prolonged unemployment.* As the "iron rice bowl" of a job for life in state firms crumbles in the face of market reforms, structural unemployment will increase and strain the unemployment insurance system.

Making labor markets flexible

There have been few formal reforms of the market for labor. But structural change and competition are profoundly affecting the way labor markets work in China, especially in rural areas. The rapid growth of jobs in township and village enterprises during the 1980s was forged largely by market forces and with little government oversight. Similar pressures are now playing an increasing role in urban markets, but the difficulties of state enterprises have made the transition more difficult.

Thus it is unlikely that China's labor market will become much more flexible unless a determined effort is made to change policies and institutions. The market could even become more segmented, with widening income inequalities across skills, regions, genders, and firms. But if policies and institutions are designed to create a flexible and integrated labor market, the outcome will be good for growth and for workers.

Easing restrictions on rural-urban migration

One good indication of how labor markets function in China is the estimated 120–140 million "surplus" workers in agriculture—roughly 35 to 40 percent of the agricultural workforce. These workers could earn significantly more in nonagricultural activities (box 4.1), but for various reasons are constrained from doing so.

Perhaps the main factor inhibiting migration to towns and cities is the urban welfare system. Migrants find it difficult to qualify for health, education, and housing facilities, which (like other benefits in urban areas) are tied to state enterprise employment. For example, state enterprises and government agencies control almost 80 percent of the urban housing stock and allocate almost all of that to their employees.

Migrants are also acutely aware that their legal status is ambiguous. As a result they often do not report unsafe conditions or illegal practices at work. Their inner-city townships are occasionally torn down, adding even greater uncertainty to their already precarious existence. They face great difficulty in formally registering as urban residents, although some cities are beginning to issue "blue cards" or temporary permits. Once these expire, however, the prospect of eking out a marginal existence in the city often encourages migrants to return to their villages.

Would-be migrants are also deterred from moving by conditions at home. For example, under the household responsibility system local authorities periodically reallocate land among households. If a family is away at the time, it may lose its allocated land. Another reason to remain on the farm is to fulfill the annual grain quota set by the authorities (although purchases from the open grain market are increasingly accepted in lieu of own production).

Despite these barriers to mobility, the stream of rural-urban migrants in the 1980s has turned into a flood. Recent estimates put the number of migrants in urban areas at around 80 million,[1] although genuinely long-term migrants are nearer to 44 million.[2] A temporary migrant typically stays in the city for about 200 days and then goes home.[3] Among longer-term migrants, peasant women in the Pearl River Delta work for as much as three to five years before returning permanently to their villages.

About two-thirds of all migrants, temporary or permanent, work in the same province as their home village.[4] Three-quarters are rural-urban migrants; the rest are rural-rural. Large, prosperous coastal cities—Beijing, Shanghai, and Tianjin—attract the largest share of immigrants, while provinces with low land-population ratios and high concentrations of state enterprises have the highest emigration rates. More than half of all migrants came from farms of less than 0.33 hectares (table 4.1).

Labor migration lubricates the rasping process of China's structural change. It is one of several human dimensions in the transition from an agricultural to an industrial economy and from a rural to an urban society. It is also an essential element in the economy's continued rapid growth. For example, the steady flow of skilled workers to urban industry keeps real wages low even as growth remains strong, supporting China's continued competitiveness in international markets. Migrant workers earn 70 to 80 percent of what their

BOX 4.1

Moving off the farm

Leaving the land can be attractive. A recent study of Shandong Province's Zouping County found that the going wage in nonagricultural activities was seven times what farmers earn. In six villages of Xiayu County in Henan Province, the returns to working more in agriculture were near zero, but significant in other activities. A survey of eight provinces found that the marginal returns from working in agriculture were considerably lower than from rural industry or migration.

Source: Cook 1996; Hare and Shukai 1996; Knight and Song 1995.

TABLE 4.1

Migrant workers usually come from small farms

Share of rural migrant labor by farm size of migrant household, 1994 (percent)

Farm size (hectares)	Share
0.00–0.33	51.1
0.34–0.66	31.4
0.67–1.00	9.7
1.00–1.33	5.0
More than 1.33	3.2

Source: Ministry of Agriculture, State Statistical Bureau (reported in Asian Development Bank 1996).

urban counterparts make, and their social overhead costs (education, health care, unemployment insurance, pension benefits) are almost nil.

More important, the opportunity to migrate gives poor rural workers and their families a chance to escape the vicious cycle of poverty. Urban migrant workers earn several times more in their new urban jobs than they would back home. In some cases their remittances average 1,200 to 1,500 yuan a year—two to three times the poverty line in poor provinces.[5] Assuming there are 40 million migrants, total remittances could amount to more than $7 billion, or 1.2 percent of GDP. The families of migrants use the money for many things, including food, fertilizer, agricultural inputs, school fees, and health charges.

Despite all these benefits, the government is concerned that uncontrolled migration could increase the size of city slums, strain the finances of city governments, and lead to social ills such as crime and prostitution. But the government also recognizes that China is set to urbanize rapidly (figure 4.1). Thus its emphasis is on managing the process, not least by supporting the development of hundreds of small towns and encouraging migrants to go to them (box 4.2).

At the same time, however, local governments in some provinces and counties are introducing regulations that discriminate against migrants. This is a disturbing development, since it could fracture China's labor market into several parts. Urban islands of formal labor markets with high wages would be surrounded by a rural sea of informal labor markets with low wages. The combination would not have an immediate, catastrophic effect on China's development. But its impact on economic performance would be felt gradually as restrictive policies sapped the market's capacity to efficiently allocate labor resources.

FIGURE 4.1

China's growing cities

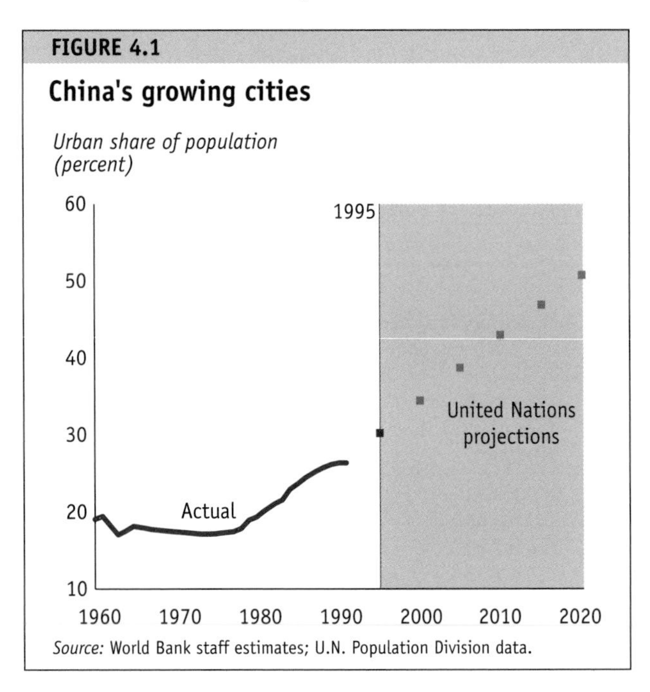

Source: World Bank staff estimates; U.N. Population Division data.

Making urban labor markets more flexible

Although migrants have spurred the development of labor markets on the fringes of cities, they have had little effect on labor markets in cities themselves. These continue to be rigid and distorted, paralyzed by the

BOX 4.2

Is small beautiful? Developing small towns to absorb rural-urban migration

"China does not just need better cities. It needs *more* cities." This remark by a senior official summarizes the government's policy of developing small towns in response to rapid urbanization and the flood of rural-urban migrants. Today 23 percent of migrants move to large cities, up from 8 percent in 1986.* This increase is beginning to strain the absorptive capacity of China's large cities. The Ninth Five-Year Plan therefore emphasizes the need to "develop a number of small towns in an orderly way, guide a small number of small towns with good conditions to grow into small cities, and guide other small towns to develop their transport facilities, improve public utilities, and protect the environment."

Studies of urbanization show that cities grow because they are economic. Infrastructure and other services are more productive when they serve concentrated populations and economic activities. Big underlying forces, such as macroeconomic and pricing policies, industrialization, and new trade opportunities, can undercut government efforts to stimulate small town development. So, if the government feels compelled to implement a small town development program, it should:

- Focus infrastructure investment on towns that are already growing fast and are located along key transport links. Market forces are probably already propelling these places to become the cities of the future.
- Develop the capability of governments in these towns to raise and allocate their own resources efficiently and independently, rather than make them rely on subsidies from Beijing or the provincial capital. Although subsidies could accelerate their development, small town development must be the sort that can be sustained without further handouts.

* Chinese Academy of Social Sciences 1997.

problems of state enterprises and government concern with rising urban unemployment. Official unemployment in urban areas climbed from 2.3 percent in 1991 to 3.0 percent at the end of September 1996.[6] But in many cities workers are retained on the rolls of enterprises even though they are not paid. If their numbers are included, government officials estimate the real unemployment rate to be close to 7.5 percent.[7] For example, in five large cities recently surveyed (Beijing, Chongqing, Guangzhou, Shanghai, and Shenyang) the unemployed and the furloughed together account for 13 percent of the labor force.[8] Even this probably understates true unemployment in urban areas, since many workers are probably not really needed in the enterprises in which they are employed. Government officials estimate that 15–20 percent of state enterprise employees could be released without affecting the output of their firms.

Many factors discourage state enterprises from laying off staff. As noted earlier, these enterprises provide their employees with various social services, including housing. Managers often need permission to lay off workers, though many formally have autonomy on hiring and firing. Local officials tend to grant such permission reluctantly because they are worried about the social implications of rising unemployment.

The missing ingredient, of course, is for workers who lose their jobs to find employment. In a dynamic economy such change is entirely normal, and the government is making arrangements to facilitate the process. The Ministry of Labor runs 2,700 retraining centers to upgrade the skills of redundant workers, and 31,000 employment service centers to help displaced workers locate jobs. The retraining programs have reached 1.2 million people already, on top of the 3.2 million who are being retrained in schools and universities. But these offices cannot cope with the demand for their services. As for the employment information offices, they are hamstrung by inadequate funding and often outdated information, are slow to respond to individual needs, and are not integrated into a citywide (let alone nationwide) network.

As more workers move into the nonstate sector, the government's concern will shift to the longer-term issue of workers' rights in a market economy. China needs to develop institutions and mechanisms for conducting wage negotiations between workers and employers. Beyond that, many nonwage issues will arise. For example, workers often encounter health risks on the job. According to the International Labour Organization, in industrial countries occupational injuries and deaths can cost as much as 4 percent of GNP. No equivalent data are available for China, but the costs are likely to be much higher. In a market economy there is a valid concern that, without government intervention, employers may not reveal the full extent of risks that workers face. Even if they do, workers may still accept dangerous conditions for fear of losing their jobs.

China has many labor laws and regulations stipulating health and safety standards in the workplace. But their enforcement, especially in the sprawling nonstate sector, is weak. Rather than more stringent laws, the emphasis should be on better enforcement and workplace design. Other measures can also be effective and cheap: for example, a government information program on the risks of toxic chemicals, or civic organizations that enforce legal standards for safety. In many other countries trade unions also play that kind of role, which makes them affordable and effective.

Indeed, the experiences of other countries hold many lessons for China to consider. First, workers bodies (such as free trade unions) are useful in organizing workers into a single group with a collective bargaining power that matches the power of employers. At their best, these organizations help balance the need for enterprises to remain competitive with the aspirations of workers for higher wages and better conditions.

Second, for market forces to work effectively in allocating and rewarding labor, it is important that enterprises operate in a competitive environment. Competition tends to force managers and unions to reach wage agreements that match productivity with pay.

Third, the best framework for achieving positive economic effects is wage bargaining between unions and managers at the enterprise level. China can learn from Hong Kong or Japan in this regard, where unions are organized along enterprise lines and operate under strong competition in product markets.

Finally, there ought to be a balance between protecting the rights of unions and limiting their potential monopoly power. Decentralized bargaining requires strong guarantees of union rights. If the system is abused by employers, industrial relations could deteriorate and worker unrest could grow.

Expanding education

Education plays a central role in the smooth functioning of labor markets. Higher literacy rates in rural areas, especially among women, spur the movement of workers from agricultural to nonagricultural activities. And the next phase of economic development will require managerial and financial skills that only a well-established tertiary education system can provide.

China's earlier investments in education paid handsome dividends as the economy was gradually liberalized. In 1990 China's illiteracy rate of 22 percent was about average relative to other countries (figure 4.2), and it has been declining steadily. In the 25–54 age group—those who went to school after 1949—illiteracy is only 16 percent, well below the 62 percent among those 55 and older. More impressive still, illiteracy among 10–24 year olds is only 5 percent (figure 4.3). As the more literate younger generations grow older, average illiteracy will continue falling. There is a danger, however, that rural children who do not move on to lower secondary school will actually *lose* literacy skills, so the government should continue its efficient and effective adult literacy program.

This creditable record has one blemish: the wide gap between female and male literacy rates (figure 4.4). Boys were given priority in education, and the gender gap began to narrow only when the male illiteracy rate had dropped to a low level. The rapid decline in female illiteracy after 1964 was the cumulative result of earlier government literacy programs targeted at women.

Although China has a wide education base, the apex of the education pyramid is much narrower than in other low-income Asian economies with much weaker education records. Nearly all Chinese children spend a relatively short period at school: 5.6 years on average, similar to that in India but well short of China's East Asian neighbors. Of 1,000 children starting school today, only 275 will stay on past lower-middle school, and almost as many will not make it past primary school (figure 4.5). The government's Ninth Five-Year Plan sets a target of achieving the norm of nine years of schooling by 2000. Reaching this target will require the participation rate in lower-middle schools to rise by 50 percent, and the dropout rate to fall. It might be helpful to eliminate examinations at the end of primary school, as has already been done in parts of China. Doing so would encourage students to enter secondary school and would benefit girls in particular, since they tend to drop out in greater numbers at this stage of their education.

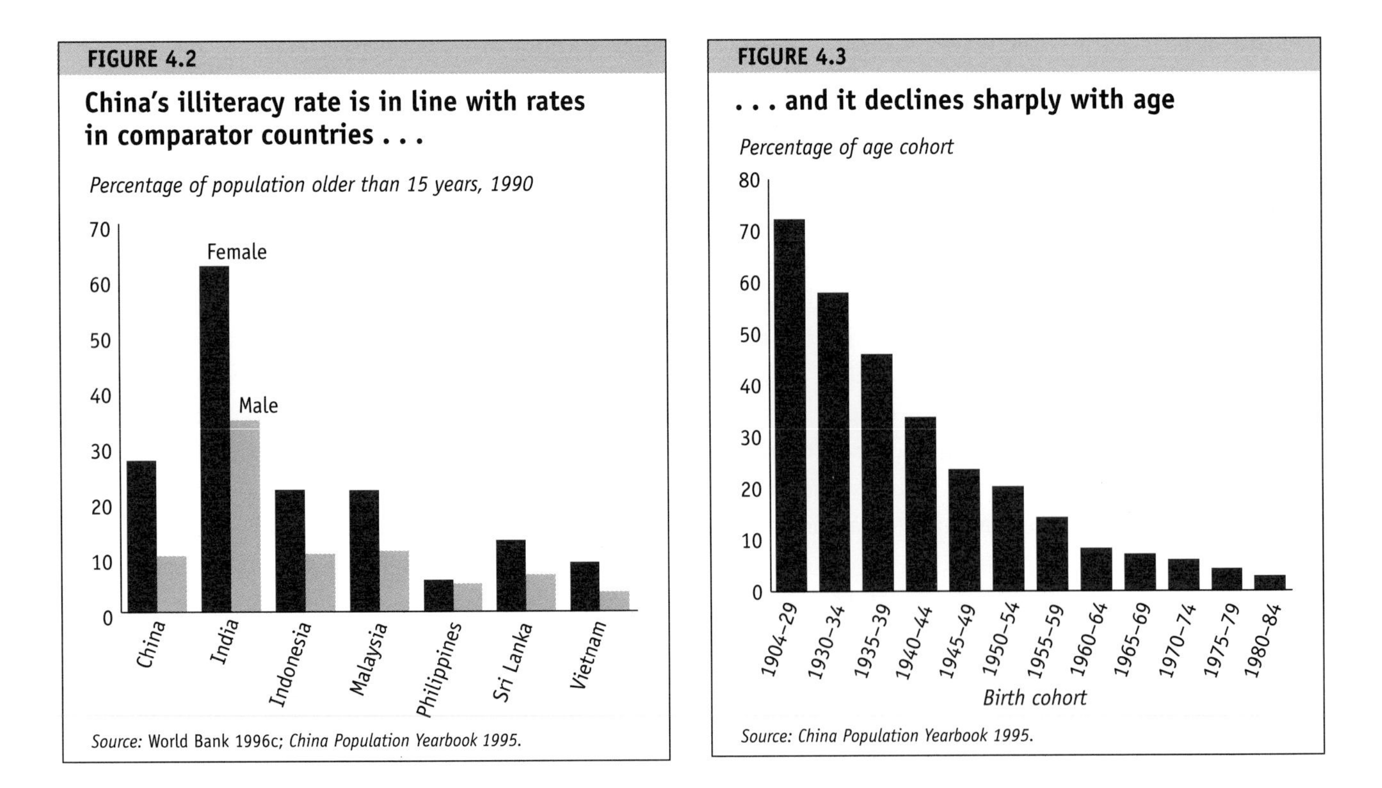

FIGURE 4.2

China's illiteracy rate is in line with rates in comparator countries . . .

Source: World Bank 1996c; *China Population Yearbook 1995.*

FIGURE 4.3

. . . and it declines sharply with age

Source: China Population Yearbook 1995.

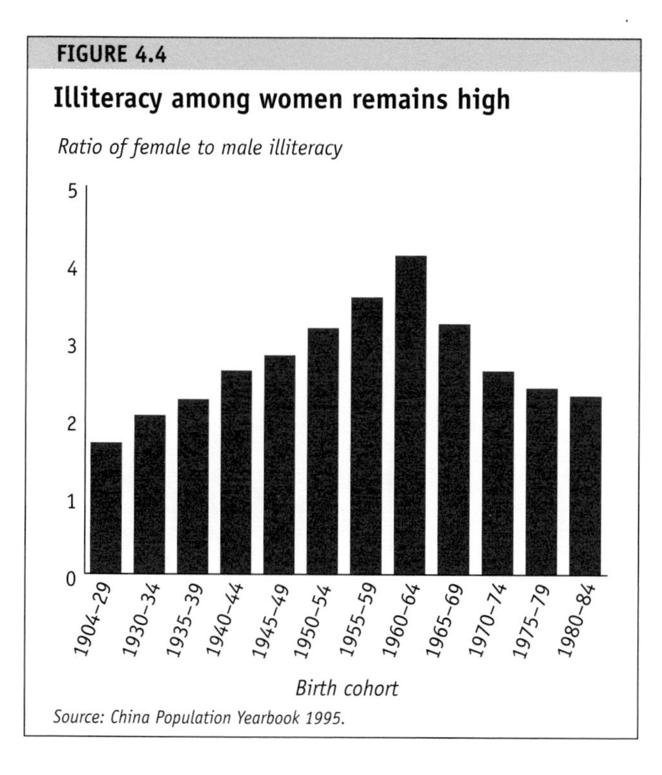

Source: China Population Yearbook 1995.

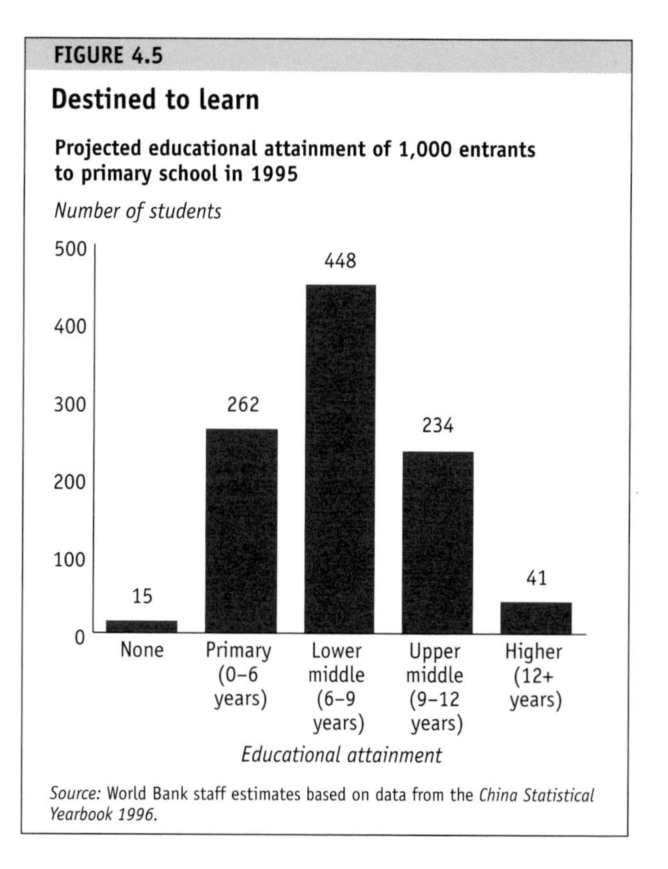

Source: World Bank staff estimates based on data from the *China Statistical Yearbook 1996.*

Upper-middle schools are not only restricted to a small portion of children, but are also heavily biased toward vocational and technical education. The emphasis on vocational and technical education in China is driven by three concerns: keeping unemployment among school leavers low through a closer fit between supply and demand, expanding secondary schooling without increasing pressure on the tertiary level, and upgrading the technical skills of the labor force.

These concerns are less compelling in a market system where demand for skills and occupational categories is changing rapidly. The new concern is not whether school leavers can slot into a job immediately, but whether they can adapt to changes in job requirements throughout their working lives. A better balance between vocational and technical education and general education has merits that are underscored by international experience. Vocational and technical education enrollments in the Republic of Korea, Malaysia, and Thailand are declining as employers and students recognize that general education equips people for the demands of a modern economy integrated with the world trading system.

As the emphasis changes, general education programs in secondary schools need to emphasize quality and creativity. The experiences of other countries suggest that a science curriculum that stresses experimentation and scientific inquiry and a mathematics curriculum that encourages problem-solving make school leavers more receptive to on-the-job training. China could benefit from a more flexible approach to curriculum development, with the State Education Commission involving employers and ministries in its plans.

As far as higher education is concerned, the echoes of the Cultural Revolution will continue to be heard for some time. During that period most universities were closed and the number of graduates declined (figure 4.6). Students who did graduate had received a poor education. Today, more than a quarter of China's graduates between the ages of 25 and 64 are older than 50 and will have retired by 2010. Merely replacing them in the workforce will require substantial increases in higher education enrollments. There is some evidence that this has been happening, especially since 1991 (figure 4.7). Even so, by 1994 only 2.4 percent of the university age cohort (18–22 years old) was enrolled in higher education.[9] That compares with 9 percent in Thailand, 10 percent in Indonesia, 20 percent in Hong Kong, 39 percent in Taiwan, China, and 51 percent in Korea.[10]

FIGURE 4.6

Educational echoes of the Cultural Revolution

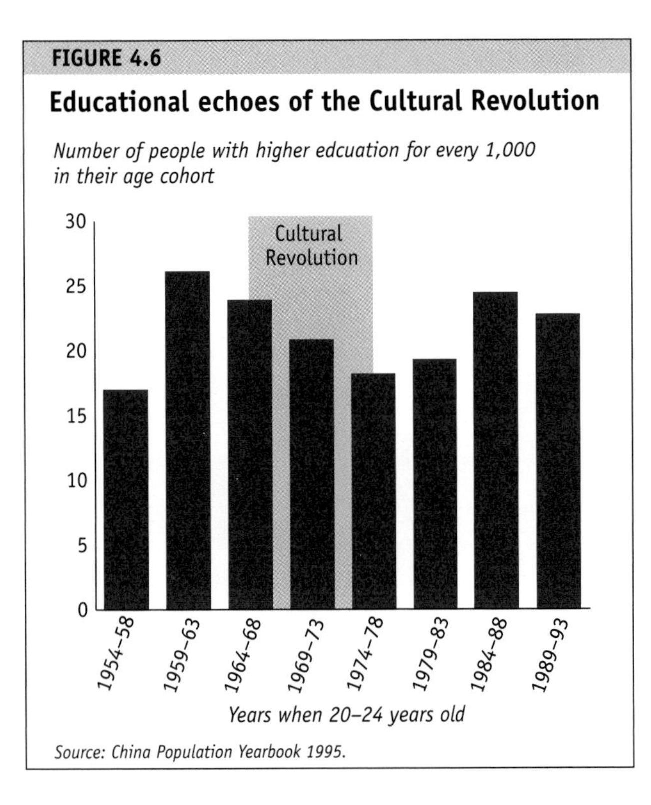

Source: China Population Yearbook 1995.

FIGURE 4.7

Enrollment in higher education has grown sharply in recent years

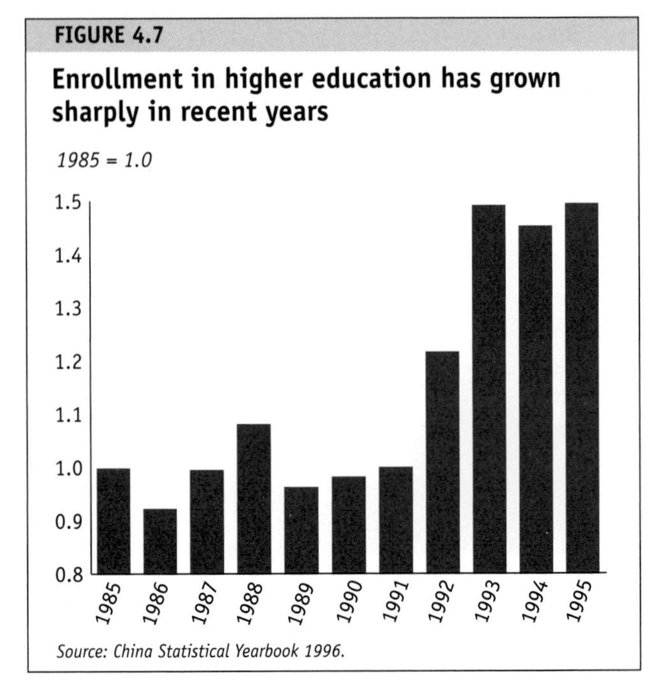

Source: China Statistical Yearbook 1996.

Inside China's higher education institutions there is a strong bias toward engineering, which accounts for 39 percent of all enrollments. This emphasis can lead to high unit costs. China's public spending per student in higher education is 175 percent of GDP per capita; the average in other East Asian economies is 98 percent. Yet evidence suggests that a significant portion of engineering students eventually work in occupations that do not require engineering training. A key challenge for the future is to design a flexible education system that supplies skills in line with changing needs and shifting employment patterns. For example, the government could place more emphasis on science, mathematics, and humanities programs to supply more qualified graduates to the growing services sector.

Although the case for new and additional expenditures on education is compelling, it is equally important to consider where these resources will come from. The government (including local government) already foots 87 percent of education expenditures; most of the rest comes from state enterprises or tuition fees. The government's budget will remain under pressure, and state enterprises are going to shed their social obligations. So the main source of extra revenues for education will have to be a combination of higher tuition fees, student loans, and financial aid programs (box 4.3).

Protecting the vulnerable

A caring society has formal mechanisms that help vulnerable people cope with structural change and market reforms. Although there are many such people in China, here we focus on five groups who are particularly vulnerable to the effects of markets.

The poor

Rapid growth has helped lift 200 million Chinese out of poverty since 1978. But most of the reduction occurred in the first six years, when agricultural incomes soared following the introduction of the household responsibility system. Progress stalled in the mid-1980s, but has picked up since 1992. In 1996 about 6 percent of the population—some 70 million people—were living below the government's absolute poverty line. Against the World Bank's international poverty line of $1 a day, however, the poor make up 22 percent of China's population (figure 4.8).[11]

In most cases the poorest people are entire communities living in isolated, upland regions of the interior with few if any natural resources. Although they have rights to land, the land is usually so poor that it is impossible to produce enough food to survive. Thus, unlike most farmers in China, these groups are net purchasers of food, and so are hurt whenever the relative

BOX 4.3

Financing additional expenditures in higher education

According to one scenario explored by the World Bank, China's expenditures on higher education could grow by 8.2 percent a year (in real terms) until 2020. This projection rests on the rather strong assumptions that unit costs can be significantly reduced (through higher staff-student ratios and more efficient use of buildings) even as the quality of education improves considerably. Under this scenario the share of public expenditures falls from 74 percent in 1994 to 51 percent in 2020, while the share of student fees rises from 9 percent to 31 percent and the share of institution-generated income from 14 percent to 18 percent.

Source: World Bank 1997a.

price of food rises. Moreover, the poor tend to be less educated, less healthy, and have more dependents for each working-age person. In several of the poorest villages at least half the boys and almost all the girls do not attend school and are unlikely ever to be literate.

Most of China's poor households lack the physical and human assets to benefit much from the mainstream forces of growth. For this reason, in 1986 the government established a Leading Group for Poverty Reduction to develop coherent policies for the poor. By 1996 the group had identified 592 poor counties (on the basis of average per capita incomes in 1992), and they have become the target for coordinated poverty reduction programs by government agencies.

As the numbers of the poor have dwindled, further reductions have been increasingly difficult to achieve. Continued progress will require more careful targeting, perhaps concentrating on individual townships and villages. That approach will help the government to include in its programs people who do not live in the designated poor counties but who account for about half of the poverty total. Careful targeting will also mean differentiating between the chronic and the transient poor. About 40 percent of poor people live in households that are not poor on average, but that have suffered a temporary decline in incomes.[12] Programs to help these households would differ significantly from those designed for the chronically poor.

Apart from better targeting, poverty programs may need other changes of emphasis. Many poor people undoubtedly benefit from investments to improve rural infrastructure, agricultural productivity, and off-farm employment. But the returns from such investments will decline once the bulk of the poor are in remote areas with few natural resources. These people will benefit from a renewed emphasis on basic health and education, combined with assistance to find employment elsewhere.

FIGURE 4.8

Sharp drop in poverty

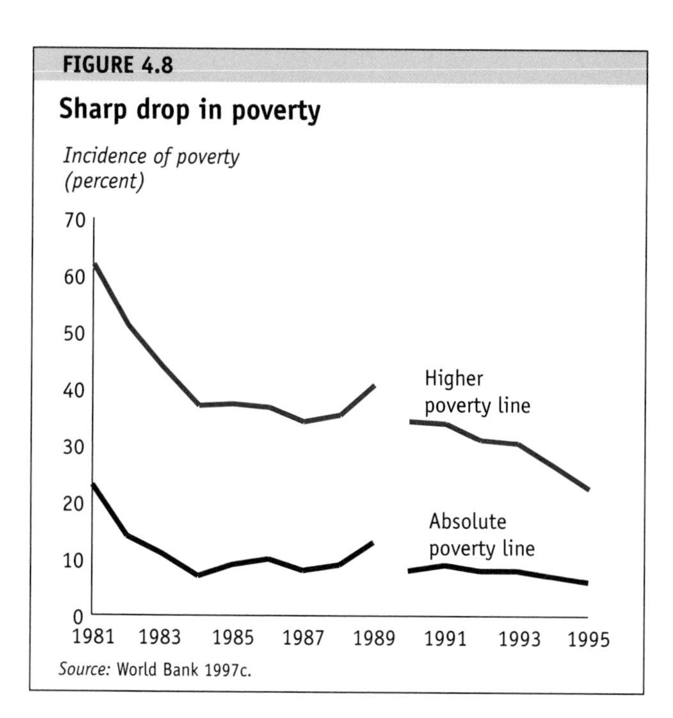

Source: World Bank 1997c.

The elderly

China is going through a profound demographic transition. A stringent birth control policy has brought an extraordinarily steep decline in the fertility rate. This drop has transformed the age structure of the population, which is now somewhere between the patterns of high-income and low-income economies (figure 4.9).

One major consequence of this change is that the elderly account for an increasing portion of the population. Today, their share is over 6 percent, but by 2020 it is expected almost to double to 11 percent (table 4.2). In today's high-income economies a similar aging took more than a century. Even by the standards of other developing economies, China's population will age very quickly. By 2020 the share of the elderly in China's population will be half as much again as in, say, India, Indonesia, Malaysia, or Vietnam.

Ensuring the financial security of the elderly will be a major challenge because labor force growth is slowing. Indeed, by 2030 the total workforce is projected to start

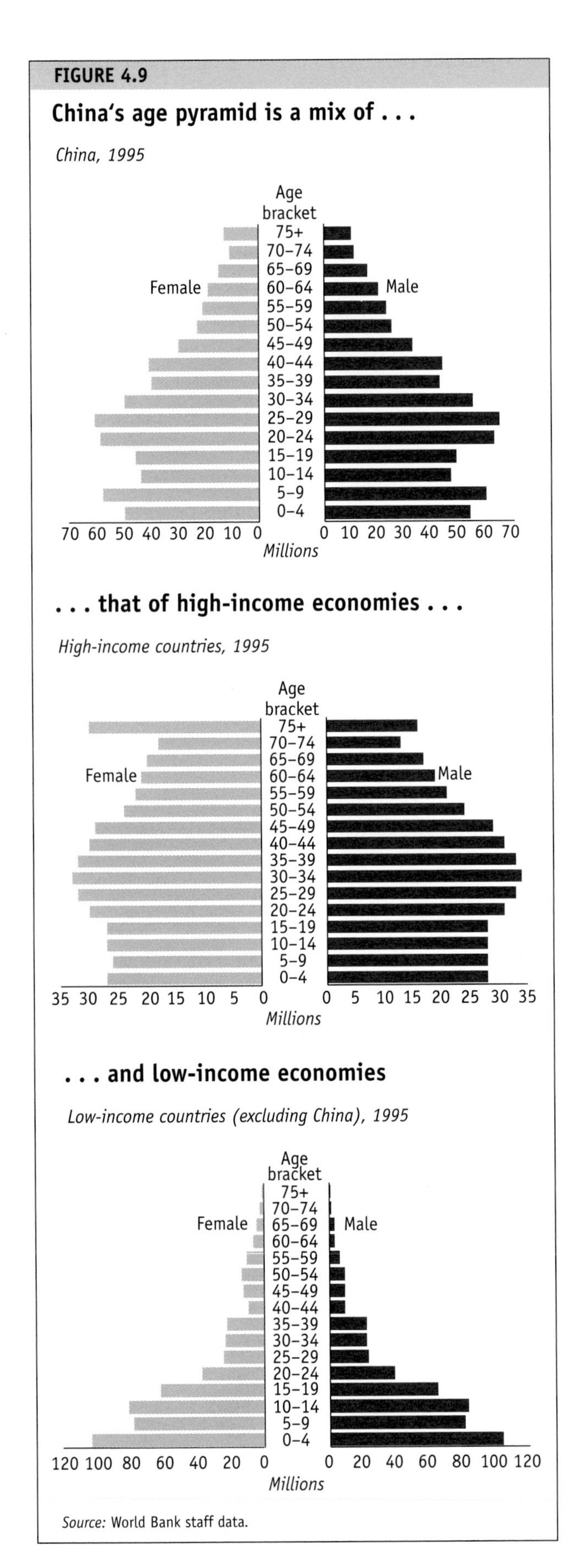

TABLE 4.2

China's demographic profile, 1995 and 2020

Indicator	1995	2020
Population (thousands)	1,200,241	1,425,288
Labor force (thousands)	811,402	987,778
Fertility rate (percent)	1.95	2.00
Life expectancy (years)	69	73
Children/population	26.0	19.9
Elderly/population	6.4	10.8
Dependency ratio[a] (percent)	47.9	40.9

a. Ratio of nonworking-age population to working-age population.
Source: World Bank data.

declining in absolute numbers. Today there are ten people of working age for every pensioner.[13] By 2020 there will be six, and by 2050 only three.

More immediately, China is already facing a pension crisis in its state enterprises. Although employment in those enterprises is growing slowly, the number of their pensioners is rising rapidly. Some enterprises have more pensioners than workers. With these firms in a weak financial position, managers occasionally have been forced to stop pension payments.

The problem lies in the pension system inherited from the era of central planning. It is a pay as you go system with defined benefits that cover mainly the retired employees of state firms in urban areas. In nonstate firms coverage is spotty, ranging from 20 percent to 90 percent (depending on the locality). By international standards the elderly receive a small portion of their income from pensions (figure 4.10).

In the past enterprises were responsible for paying their own pensioners. But in some places this responsibility has recently shifted to local governments, who pay pensions from pooled funds that are replenished regularly by enterprise contributions. Today some 110 million workers are covered by this pooling arrangement, which has helped spread risks and is especially useful for firms with large pension bills.

Even so, the system is untenable. Firms with more pensioners still make a larger contribution to the pension pool and are still responsible for most of the payments administration and the health and housing needs of pensioners. This arrangement makes it difficult to liquidate bankrupt firms. Moreover, viable firms in places with a large concentration of pensioners could find their competitiveness eroded because of their larger contributions to the pool.

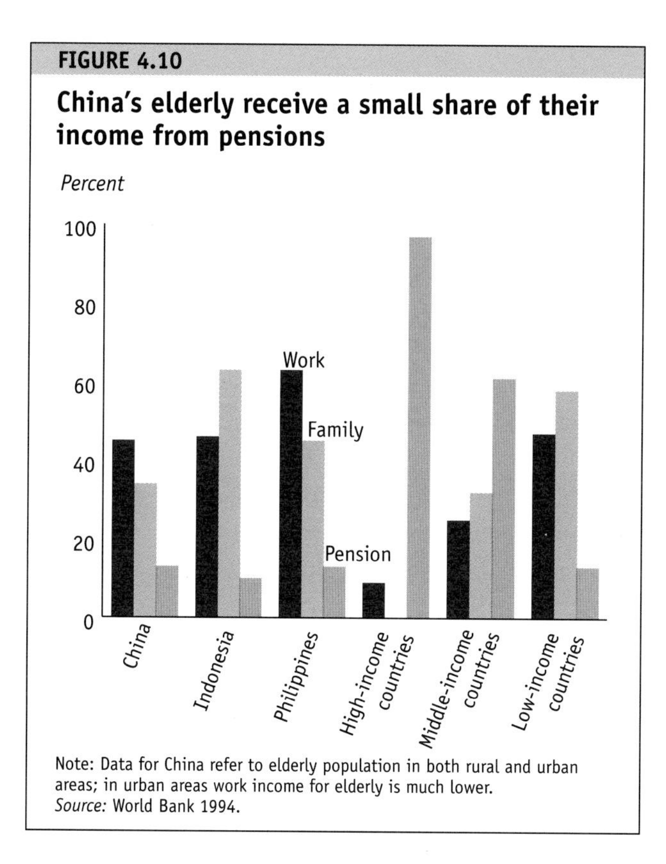

FIGURE 4.10

China's elderly receive a small share of their income from pensions

Note: Data for China refer to elderly population in both rural and urban areas; in urban areas work income for elderly is much lower.
Source: World Bank 1994.

Most important, the pooling system fails to make pension benefits portable from one locality to another, hampering labor mobility and impeding state enterprise reform. Individual accounts are only notional, not real. Worker contributions are used to pay existing pensioners, rather than accumulating in workers' individual accounts. Unless this system is reformed, the contribution rate will have to rise steadily. By 2035 the average rate would pass 40 percent, making most enterprises financially nonviable.

In 1995 the government introduced a master plan that aims at four "unifications" by 2000: a unified system that treats all enterprises and workers equally, with unified standards, unified management, and unified use of funds. However, the government proposed two models—one emphasizing individual accounts and the other a larger social component—and allowed local authorities to introduce schemes using any combination of features of the two plans. The result is hundreds of incompatible schemes across the country.

The World Bank, working with government experts, has since proposed a three-pillared system that combines social pooling with funded individual accounts.[14] The first pillar would provide a basic pension to keep retirees above the poverty line, with the same basic pension graded only by length of service. After forty years of service, for example, retiring workers would obtain 24 percent of their last wage packet. A contribution rate of about 9 percent of enterprises' wage bill would be sufficient to finance these payments.

The second pillar would comprise mandatory, fully funded individual accounts paid for by an 8 percent contribution rate split equally between workers and enterprises. Assuming that the rate of return on pension funds equals the rate of growth in wages, the pensions from these accounts would be equivalent to 35 percent of workers' final wages.

The third pillar would be a supplementary pension, which employers could choose to provide or individuals could choose to save for. Pension funds and insurance companies would offer these accounts, much as they do in other market economies.

This scheme reduces the pension burden in three ways. First, it assumes that the retirement age is gradually increased to a unified 65 years (from 60 for men and 55 for women). Second, it indexes pensions to prices, not wages. Third, the first two pillars together result in a wage replacement rate of about 60 percent.[15] Although lower than the current wage replacement rate of more than 80 percent, this level is much closer to replacement rates in most other countries (40–60 percent).

Even if all these reforms are adopted, finance is still needed for the implicit pension debt of state enterprises—that is, pensions already being paid plus the accumulated pension rights of workers under the old system. This debt is about 50 percent of GDP, much less than in many other developing countries. There are many options for financing it, including an additional contribution from state employees (who comprise almost all the beneficiaries), bonds issued by the government, and sale of state enterprise assets. Each option has its own constraints and advantages, and some combination of all three could be used.

Whichever reform is chosen, it will be more effective if it is begun soon. Today's rapid growth and high savings will make it easier to absorb the incremental costs of reform. Reforms will get harder over time as more workers move from agricultural to nonagricultural activities. Equally, pension reform is a necessary element in reforming state enterprises: delaying one will inevitably delay the other. And more positively, the development of a savings-based pension system will stimulate the rapid growth of capital markets.

Still, these reforms do carry risks, with much depending on the rate of return earned on the accumulating funds. For the first pillar, for example, each percentage point reduction in the return on invested balances would require the contribution rate from enterprises to be an extra 0.6 percent of wages. If the future real return on invested balances were no higher than the real return on deposits over the past five years (0.4 percent as opposed to the 4.0 percent assumed by the simulations), the contribution rate would have to be about 2.2 percentage points higher (figure 4.11). The higher is the contribution rate, the greater is the probability of evasion, especially by nonstate firms.

Achieving high returns on pension funds will require a new approach in China, since the experiences of other countries show that privately managed funds earn better returns than publicly managed ones (figure 4.12). How such arrangements are put in place requires careful consideration, because the funds will be managing huge sums. By 2030 their accumulated surplus could reach 13 trillion yuan, or $1.6 trillion (in 1994 prices). As a share of GDP this would comfortably exceed the 1991 pension assets of Germany and Japan. Annual inflows of new money could reach 500 billion yuan ($60 billion; figure 4.13), so China's pension funds will be a major force in international capital markets (figure 4.14).

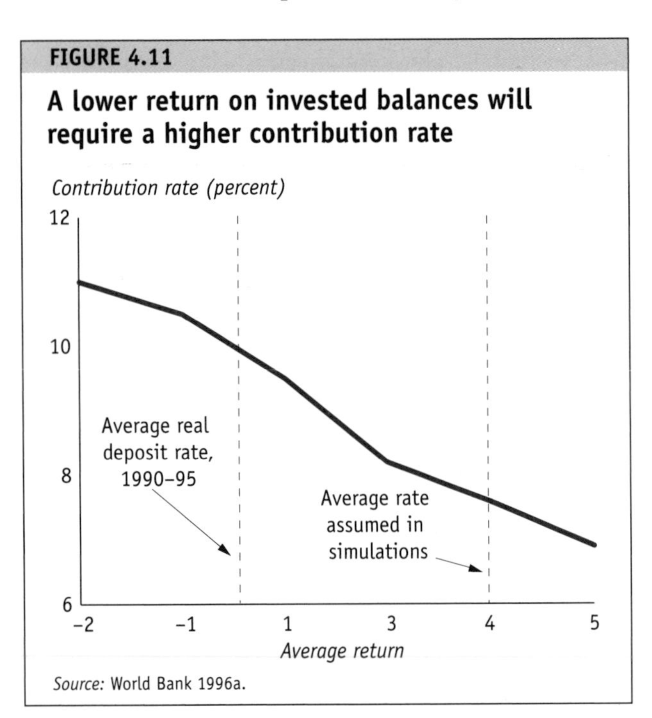

FIGURE 4.11

A lower return on invested balances will require a higher contribution rate

Source: World Bank 1996a.

The sick

Health care spending has been growing steadily in China and is expected to rise even faster over the next twenty-five years. Today China spends only 3.8 percent of GDP on health, so it will still take years to catch up with Thailand (5.3 percent), Korea (5.4 percent), Chile (6.5 percent), Colombia (7.4 percent), and industrial market economies (9.2 percent).[16]

But personal spending on health has grown rapidly since reforms began (table 4.3). In rural areas the health insurance implicit in the commune system has crumbled (see chapters 1 and 3). In urban areas state enterprises are withdrawing from providing lifelong subsidized medical care to workers and their families. In a recent survey of urban workers, concern about mounting health care costs ranked ahead of fears of unemployment and insecurity in old-age pensions. But the ramifications go beyond personal anxiety, because many cross-country studies confirm a positive link between public spending on health and labor productivity.

The government could have compensated for these reform-induced reductions in public health services by

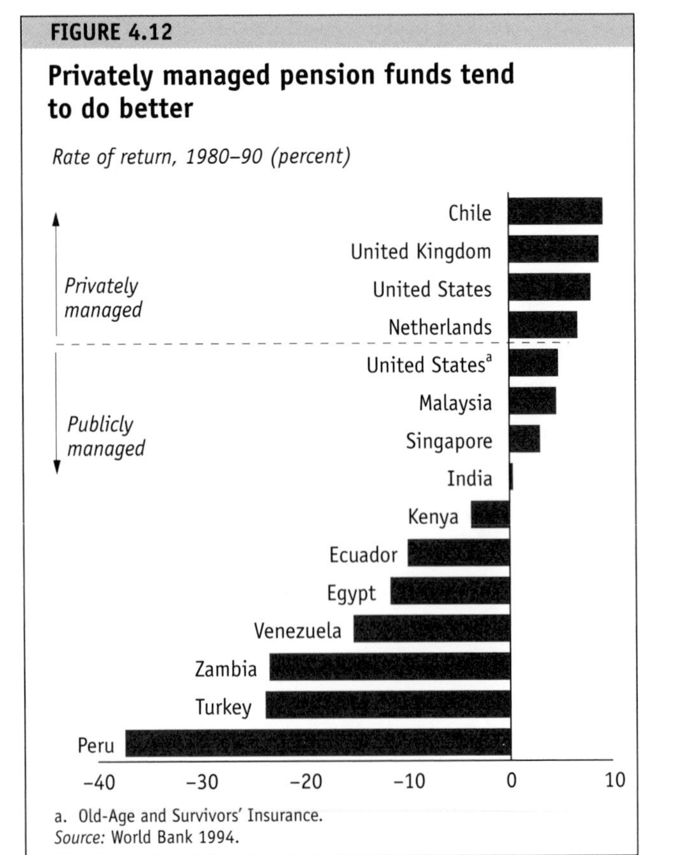

FIGURE 4.12

Privately managed pension funds tend to do better

a. Old-Age and Survivors' Insurance.
Source: World Bank 1994.

FIGURE 4.13

Growing surpluses in China's pension funds

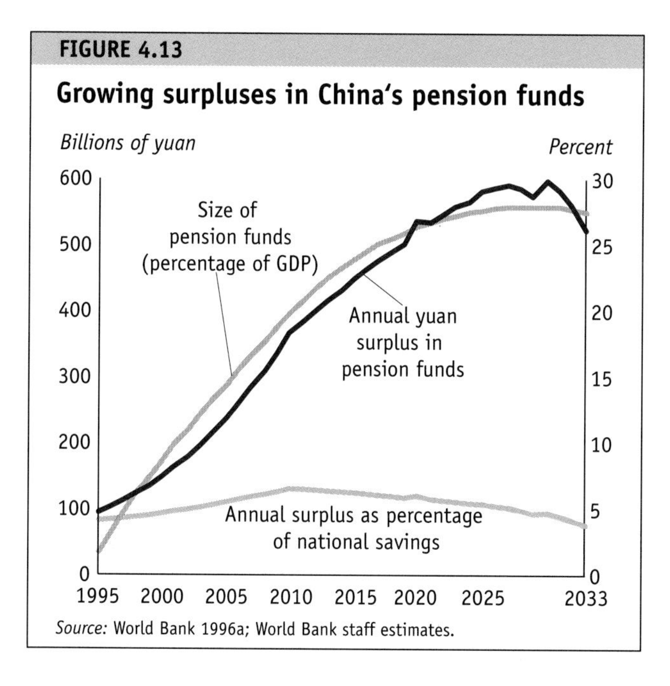

Source: World Bank 1996a; World Bank staff estimates.

FIGURE 4.14

China's pension fund surpluses will be an important world source of long-term finance

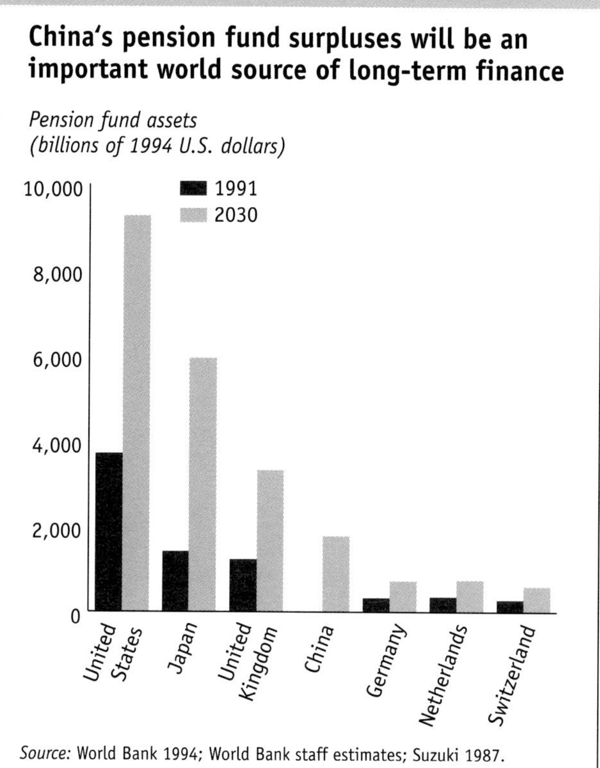

Source: World Bank 1994; World Bank staff estimates; Suzuki 1987.

increasing its own spending. Instead, its share of health spending has declined significantly, and what little extra it has spent has been mostly on family planning and civil servants. Moreover, since 96 percent of current health expenditures comes from local governments, the distribution of spending varies enormously across provinces and between urban and rural areas. For every 1,000 people, there are 75 percent fewer hospital beds in rural than in urban areas, 75 percent fewer doctors, and 80 percent fewer nurses. Per person, rural areas receive a fifth of what urban areas get in health subsidies. Since rural areas are barely covered by health insurance schemes, most services there are fee-based. The sick often must buy medical treatment at the cost of slipping into poverty—or, worse, forgo treatment but pay the price in ill health or death.

Although China's health indicators are above the norm for countries at its level of development (figure 4.15), recent trends have not been encouraging. Surveys and censuses indicate that the under-5 mortality rate stopped falling after 1985 and may have risen since.[17] China has one of the world's highest incidences of viral hepatitis. A 1990 survey of seven provinces found that 7 in 1,000 people in rural areas suffered from pulmonary tuberculosis. Predictably, the poor are particularly at risk. A study of thirty poor counties found that infant mortality rose from about 50 per 1,000 live births in the late 1970s to 72 per 1,000 in the late 1980s. A survey of nine provinces found that in rural areas stunting among children (low height for age) increased between 1987 and 1992.

Thus improving the health care system is an urgent challenge. There are four priorities for government action. First, target health spending to where it will bring the most benefits. There would be no better start than to focus on the 592 counties designated as poor by the Leading Group on Poverty. Beyond that, public financing of the epidemic prevention service should be

TABLE 4.3

Share of total health expenditures covered by insurance or government budget, 1978 and 1993
(percent)

Coverage	1978	1993
Government budget[a]	28	14
Urban insurance schemes[b]	30	36
Rural insurance schemes	20	2
Out of pocket[c]	20	42
Other	2	6
Memo item		
Health expenditure (percent of GDP)	3.0	3.8

a. Excludes insurance expenses for government employees.
b. Includes insurance expenses for government and urban labor.
c. Includes user fees, copayment, and fee for service.
Source: World Bank 1997b.

FIGURE 4.15

Health indicators in China are better than international norms

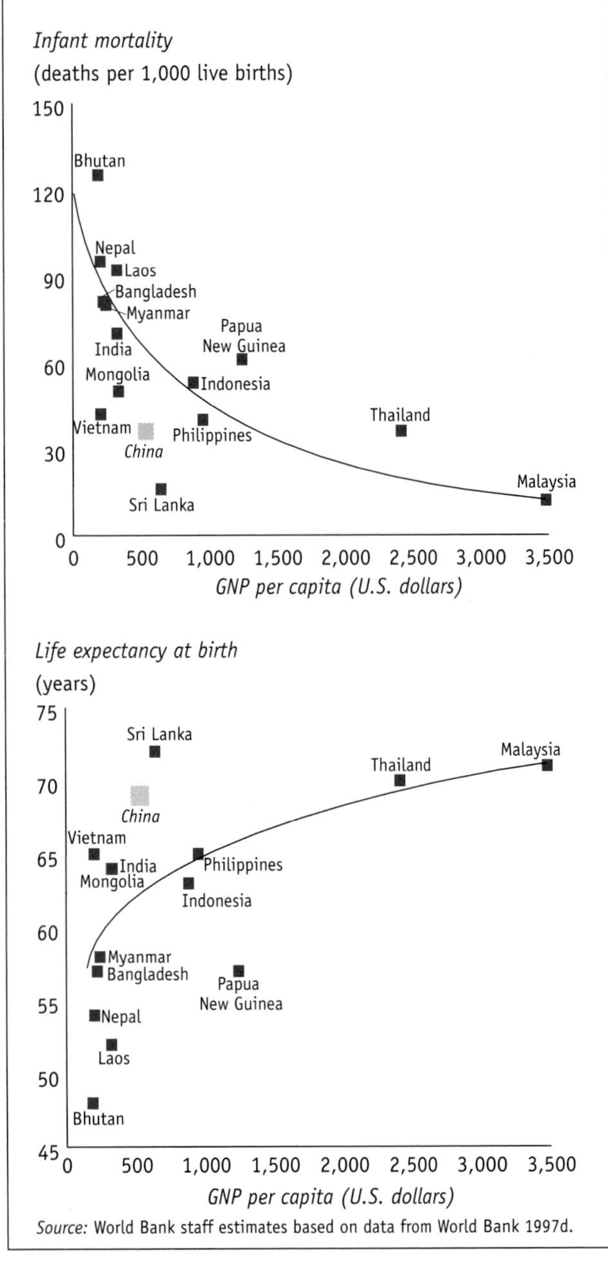

Source: World Bank staff estimates based on data from World Bank 1997d.

BOX 4.4

China's emerging health priorities

In the future, noncommunicable diseases will be the biggest killers in China. For people over 65, cancer and cardiovascular diseases will become bigger causes of death. For working-age populations, injuries and sexually transmitted diseases are expected to emerge as important health problems. The incidence of depression is also expected to grow, especially among women. In 1990 more than 180,000 women committed suicide in China, many of them aged 25–34. These trends suggest the need for government campaigns to promote safety at work and on roads, healthier diets, safe sex, less smoking (which can be encouraged further with the introduction of a tobacco tax), and long-term care for the old.

restored; it has stayed flat in nominal terms since 1986, falling by 60 percent as a share of GDP. Immunization against infectious and parasitic diseases also deserves special attention. Finally, much could be achieved through campaigns on noncommunicable diseases and injuries, with a special emphasis on tobacco control (box 4.4).

Second, align the price structure of health services more closely with costs. The Price Commission currently sets the price of essential medical services below cost and the price of high-technology diagnostic services above cost, while the State Pharmaceutical Agency allows a 15 percent markup on drugs at the wholesale and retail levels. As a result health care providers have large incentives to buy and use high-technology equipment, overprescribe drugs, and require side payments for essential health services. The pricing system should be overhauled, with gradual lifting of controls in urban areas. In the countryside controlling prices is difficult anyway, so the government should formally recognize a free rural market for health services.

Third, as the Health Policy Conference in 1996 concluded, health insurance programs need to be revamped. Two kinds currently exist—enterprise-based health care and the scheme for government employees. The enterprise-based system is springing leaks and sinking slowly, with many enterprises delaying reimbursements of health costs or lowering their contributions. Piecemeal reforms in several cities have included selective pooling of major medical costs across enterprises, transferring some risks to beneficiaries, and mechanisms for containing costs. The government has accumulated enough experience to consider expanding urban insurance reform to fifty-seven cities. Comprehensive reforms in these cities should aim to delink service provision from financing and contain costs. This calls for the introduction of individual medical savings accounts to pay for low-cost health problems, while focusing insurance on coverage of catastrophic risks and broadening the collective risk pool across enterprises and work units.

Fourth, rural health insurance programs need to be rebuilt. The Sichuan Rural Health Experiment, organized in 1989–90, is a possible guide. It covers twenty-six villages and incorporates various innovations: premiums set at 1.5 percent of average income, voluntary participation, varying reimbursement rates for inpatient and outpatient care, and low administrative costs. In most poor rural areas, however, government subsidies may be needed. Subsidies would be justified, provided the schemes incorporate lessons from the Sichuan model and others that have since improved on its design.

Women

Mao observed that "women hold up half the sky." Particularly in urban areas, women in China have an economic and social status seldom seen in other developing countries. The female labor force participation rate is 80 percent, well above the (unweighted) average of 50 percent for all of East Asia. On average Chinese women earn between 80 and 90 percent of what men earn, much higher than the worldwide average of 60 to 70 percent.

Chinese women are, however, vulnerable to discrimination and unfair treatment in the workplace. Numerous accounts suggest that they may be disproportionately affected by the restructuring of state enterprises—a trend that is likely to accelerate as enterprises withdraw support services such as child and health care. Eastern Europe's experience lends support to this view; participation rates for women declined rapidly there after market reforms were introduced.

Some labor market policies also place women at a disadvantage.[18] The most glaring is that their retirement age (55) is lower than men's (60). Some women are even required to bequeath their jobs to their children to help reduce youth unemployment. Other regulations unintentionally lead to discrimination in the workplace, especially in state enterprises. For example, the legal requirement to grant women maternity leave and child care benefits makes them more expensive to employ, so firms turn them away.

Although wage differences between men and women are small relative to differentials in other countries (table 4.4), they are remarkably durable (and tend to be higher in nonstate firms than state enterprises). They can be explained only partly by differences in such factors as schooling, on-the-job experience, and firm tenure. The main element is usually unobserved factors—which probably means discrimination.[19]

Moreover, the gap between the wages of men and women could grow for two reasons. First, the expansion of nonstate employment would boost the average wage differential. Second, reforms tend to heighten the importance of education; since men are better educated, their wages are likely to rise faster than women's. Thus reforms and growth, while raising the average standard of living for everybody, may not be good for the *relative* position of Chinese women.[20] However, much will depend on the pattern of development, the demand for women workers, and the nature of government policies. For example, if foreign investment is concentrated in labor-intensive exporting industries (which employ many women), the wage differential could even narrow.

If women's relative earnings do deteriorate, it could have profound long-term repercussions for growth. It will inevitably lead to less female participation in the labor force and reduce investment in girls' schooling. This would be serious because the education level of mothers has a stronger positive impact on the education of children than the education of fathers. Similarly, money in the hands of mothers is more likely to be spent on the health and welfare of children than money controlled by fathers.[21]

Yet mandating equal pay for women could have the opposite effect of what was intended: it could encourage noncompliance and bias employers against hiring women. Instead, the government could use general revenues to reimburse firms that offer maternity and child care benefits, treat these costs as tax deductible for

TABLE 4.4
Women's earnings as a share of men's earnings in manufacturing
(percent)

Economy	1980s	1990s
China	85	86
Singapore	62	71
Taiwan, China	66	61
Korea, Rep. of	46	54
Malaysia	73	—
Philippines	62	—
Thailand	70	—
Japan	53	50

Source: Meng 1996.

firms, or develop mechanisms for women that would allow them to pay for these benefits either in the form of reduced benefits in other areas or lower wages.

The unemployed

Unlike the pension scheme and health insurance, the unemployment insurance scheme has no major design flaws. Introduced in 1986, it is still mostly confined to employees of state-owned enterprises.[22] All firms in the scheme pay a maximum of 1 percent of their wage bill into an unemployment insurance fund, which in every case until recently has been in surplus. Benefits go to those laid off because of bankruptcy, restructuring, or the expiration of their contracts. Earlier, few workers were laid off because of bankruptcies and restructuring, so the scheme involved mainly contract workers, who now account for a quarter of the workforce in state and urban collective enterprises. But with the sharp rise in unemployment since 1993, the contribution rate is no longer adequate and will be raised shortly.

Unemployment benefits range from 50 to 75 percent of the average basic wage in the last two years of employment. Benefits last for a maximum of two years. Recipients continue to receive health benefits, but on a cost-sharing basis, and they are allowed to remain in their homes. The scheme is run by provincial and municipal governments, not enterprises. These schemes also offer job placement and training and facilitate rural-urban migration.

The scheme is not an answer to the problem of surplus labor in state enterprises (see chapter 3). Most of these workers are permanent employees recruited before 1986, when the labor contracting system was introduced. Many are relatively old and unskilled, so they have limited prospects of finding comparable remuneration elsewhere. They are more likely to join the ranks of the long-term unemployed, for whom unemployment schemes in general can do little.

Solving the problem of disguised unemployment will require a combination of measures. The government has already adopted two—expanding retraining schemes for the unemployed and lowering the retirement age. The latter reduces the burden of surplus labor borne by enterprises and exchanges it for the smaller expense of pensions. But it has complicated pension reforms; if anything, the retirement age should be raised to 65 years for both men and women. The government could also offer redundancy payments to help enterprises that restructure their operations. It could afford to do so, provided the program is staggered over the next five years. The key ingredients are that unskilled workers are covered and that there are built-in incentives for people to retrain and seek new employment.

Notes

1. Cook (1996).
2. Naughton (1996).
3. Hare (1996).
4. Li (1995).
5. Yang and Zhou (1996). In addition, World Bank missions reviewing the World Bank–financed South West Poverty Alleviation Project have observed migrant remittances averaging 1,200 yuan and often reaching 2,500 yuan a year.
6. *Oxford Analytica Asia Pacific Daily* brief, December 20, 1996, updated January 2, 1997.
7. By the end of 1996 there were 5.5 million workers officially unemployed, an estimated 9 million furloughed, and 11 million with wage arrears.
8. World Bank (1997c).
9. It should be noted, however, that almost as many students enter adult education and professional education institutions.
10. World Bank (1997a).
11. The government calculates the absolute poverty line based on an intake of 2,150 calories a day and an approximation of the cost of nonfood subsistence items based on their share in low-income household budgets. The higher poverty line is based on the World Bank's definition of $1 a day in 1985 prices adjusted for purchasing power parity. For more details, see World Bank (1996c). In 1990 Chinese household surveys introduced a more accurate pricing convention, resulting in a discontinuous series.
12. Such transient poverty constitutes about half of the mean squared poverty gap. The poverty gap measures the income transfer needed to raise the income of a poor person exactly to the poverty line. The mean squared poverty gap is the mean of the sum of the squares of the poverty gap for all poor individuals.
13. People of working age are considered those between the ages of 15 and 64.
14. World Bank (1996a).
15. The "replacement rate" is the ratio of the payable pension to the worker's wage in the final years of employment.
16. World Bank (1997d).
17. This assessment is based on a report commissioned by the World Bank (1997b). Chinese researchers, however, question the reliability of the estimates in this report, which is based on census and fertility surveys, and consider death registration data more reliable. But it should be noted that international demographers do not commonly use death registration data to estimate child mortality in developing countries, preferring instead to derive estimates from censuses and surveys. Based on death registration data, the under-5 mortality rate *declined* in the 1990s—from 61 per 1,000 in 1991 to 51 per 1,000 in 1995.
18. According to the All-China Federation of Trade Unions, a survey of 1,175 enterprises in 1993 found that 60 percent of all laid-off workers were women (Li and Hong 1996). A 1995 survey by the

Women's Research Institute of the Chinese Academy of Management Science found that women constituted 70 percent of persons fired or likely to be fired as a result of restructuring unprofitable enterprises.

19. Bauer, Feng, Riley, and Zhou (1992); Meng and Miller (1995); Yang and Zax (1996).

20. Yang and Zax (1996).

21. Agrawal and Walton (1996).

22. In some places, such as Shanghai, unemployment insurance schemes also cover collectively owned firms and foreign joint ventures.

References

Agrawal, N., and M. Walton. 1996. "Women at Work in East Asia: Does Stellar Growth Meet the Needs of Women? Should Governments Do More?" World Bank, Washington D.C.

Asian Development Bank. 1996. *Country Economic Review; People's Republic of China*. Manila.

Bauer, J., W. Feng, N. E. Riley, and X. Zhou. 1992. "Gender Inequality in Urban China: Education and Employment." *Modern China* 18(3):333–370

China State Statistical Bureau. 1995. *China Population Yearbook 1995*. Beijing.

———. 1996. *China Statistical Yearbook 1996*. Beijing.

Chinese Academy of Social Sciences. 1997. "Trend Analysis: Studies of Labor Mobility in 100 Chinese Villages." Study financed by the Ford Foundation. Beijing.

Cook, S. 1996. "Surplus Labor and Productivity in Chinese Agriculture: Evidence from Household Survey Data." Institute of Development Studies, Sussex.

Hare, D. 1996."Efficiency Considerations of Out-Migration From Rural China." Australian National University, Canberra.

Hare, D., and Z. Shukai. 1996. "Labor Migration as a Rural Development Strategy: A View from the Migration Origin." Presented at the International Conference on Rural Labor, June 25–27, Beijing.

Knight, J. and L. Song. 1995. "Towards a Labor Market in China." Background paper for *World Development Report 1995: Workers in an Integrating World*. World Bank, World Development Report Office, Washington, D.C.

Li, Q. and D. Hong. 1996. "Urban Poverty Issues and Options in China." *Population Research* 20(5): 39–42.

Li, S. 1995. "Population Mobility and Urban and Rural Development in Mainland China." *Issues and Studies: A Journal of Chinese Studies and International Affairs* 31 (September):37–54.

Meng, X. 1996. "The Economic Position of Women in Asia." *Asian Pacific Economic Literature* 10(May): 23–41.

Meng, X., and P. Miller. 1995. "Occupational Segregation and Its Impact on Gender Wage Discrimination in China's Rural Industrial Sector." *Oxford Economic Papers* 47(1):136–55.

Naughton, B. 1996. "Cities." Background paper for this report. World Bank, China and Mongolia Department, Washington, D.C.

Suzuki, Y. 1987. *The Japanese Financial System*. New York: Oxford University Press.

World Bank. 1994. *Averting the Old Age Crisis: Policies to Protect the Old and Promote Growth*. New York: Oxford University Press.

———. 1996a. "China: Pension System Reform." Report 15121-CHA. China and Mongolia Department, Washington D.C.

———. 1996b. "Poverty in China: What Do the Numbers Say?" China and Mongolia Department, Washington, D.C.

———. 1996c. *World Development Report 1996: From Plan to Market*. New York: Oxford University Press.

———. 1997a. *China: Higher Education Reform*. A World Bank Country Study. Washington D.C.

———. 1997b. *Financing Health Care: Issues and Options for China*. Washington, D.C.

———. 1997c. *Sharing Rising Incomes: Disparities in China*. Washington, D.C.

———. 1997d. *World Development Indicators 1997*. Washington D.C.

Yang, D.T., and H. Zhou. 1996. "Rural-Urban Disparity and Sectoral Labor Allocation in China." Duke University, Department of Economics, Durham, N.C.

Yang, L., and J. S. Zax. 1996. "Compensation for Holding up Half the Sky: Gender-linked Income Differences in Urban China." University of Colorado, Department of Economics, Boulder.

chapter five

Feeding the People

In recent years there has been much concern about China's future grain supplies and its ability to feed itself.[1] Some analysis projects dependence on grain imports to rise at an alarmingly fast rate.[2] At the other extreme, China is viewed as capable of supplying most of the grain it needs through the next quarter-century.[3] Which assessment is correct? The answer is clearly crucial, and not just for China. China accounts for a large share of global grain production and consumption. Changes in its net trade could move markets and grain prices everywhere.

Until recently the availability of grain was not a major concern. China's capacity to increase production in line with rapid growth in demand reflected its investments in agricultural research in the 1960s and 1970s, more fertilizer production, and more irrigation. But in all these areas progress has slowed in the past two decades. At the same time, the productivity surge following the initial agricultural reforms of 1978–81 has abated as further reforms have slowed. Although grain procurement prices recently have been raised closer to international levels, this move was not accompanied by reforms in grain marketing and distribution.

The government's Ninth Five-Year Plan and the Fifteen-Year Perspective Plan place a high priority on developing agriculture in general and grain production in particular. Three reasons motivate its concern. First, despite significant structural change, agriculture will remain an important part of the economy. Rapid economic growth will be difficult unless it is underpinned by sustainable increases in agricultural output. Second, agricultural growth will improve conditions for the poor in rural areas, and so will have big implications for income distribution. And third, the Chinese authorities are determined to avoid heavy dependence on imported grain. They fear that imports could be disrupted by uncertainty and volatility in world markets, and the possibility of trade friction with large grain exporters.

The government is right to give agriculture high priority. But there are two routes it could take. One is to pursue grain self-sufficiency even if it means taxing farmers (through production quotas and low procurement prices) or consumers (through import protection), or some combination of both. In the long run this route would be costly and unsustainable, inhibiting structural change and dampening competition and innovation.

The other route would emphasize China's comparative advantage, using trade in agricultural products as a disciplinary device to encourage efficient domestic production. This path is more likely to lead to adequate grain supplies and sustainable agricultural development, without sacrificing growth or structural change. Concerns about import dependency and volatility in international markets can be addressed using innovative, market-friendly solutions.

To illustrate the possibilities, the World Bank has constructed a model that shows demand for foodgrain rising to almost 700 million tons in 2020, up from 437 million tons in 1996.[4] Domestic production of foodgrain could range from about 600 million to 670 million tons, depending on the success of market reforms and public investments. Thus imports will range from about 30 million to 90 million tons.

These projections suggest that doomsday scenarios about Chinese agriculture are not *probable* (although anything is *possible*). Even if China were to import 90 million tons of foodgrain by 2020, these imports would rise only gradually, allowing time for world supplies to adjust and world prices to rise only marginally. For China those imports would require only 1–2 percent of export earnings in 2020 (at constant prices).

Removing constraints to domestic production

For four decades, apart from a brief period in the 1950s, China's grain production has kept ahead of population growth (figure 5.1). Between 1978 and 1984 growth was significantly above trend, due largely to the introduction of the household responsibility system. Since then, however, growth has slowed because of shortcomings in five main areas: reforms in the market for foodgrains, institutional change in the market for fertilizer, investment in agricultural research, investment in irrigation, and land reclamation and development. Each is considered in turn.

Encouraging the market for foodgrains

Although the government has freed many markets, it still controls foodgrains. It does so to hold large grain reserves (considered important for low and stable food prices) and to maintain 95 percent self-sufficiency in grain.[5] To maintain its reserves, the government buys

FIGURE 5.1

China's grain production has kept ahead of population growth

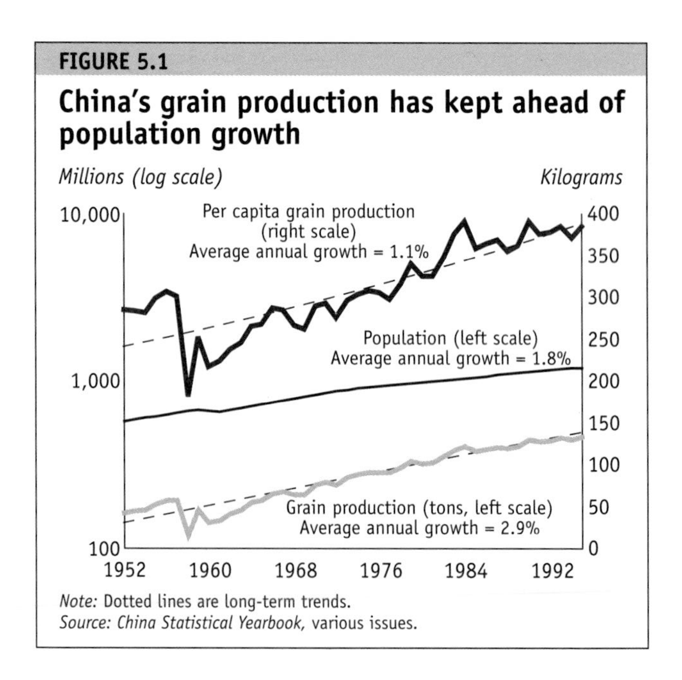

Note: Dotted lines are long-term trends.
Source: China Statistical Yearbook, various issues.

about three-quarters of all marketed grain at well below the free market price (figure 5.2). In 1996 provinces were allowed to buy their own grain under the new "governors responsibility system." Provinces with surplus grain are required to hold a three-month supply of grain, and provinces with a deficit, a six-month supply. Government bodies that buy and distribute grain are expected to operate profitably, but are subsidized through the budget and the banking system.[6]

These policies are designed to hold down the cost of grain to consumers, most of whom live in urban areas. Taken together, they cost 85 billion yuan a year—more than five times the amount spent on poverty alleviation under the National 8–7 Poverty Reduction Plan.[7] On their own the below-market procurement prices are equivalent to a 3 percent income tax on every man, woman, and child in rural areas.

The proportion of urban consumers frequenting state grain stores has been shrinking in recent years because better grain is available in the open market. Some observers have suggested that the state grain enterprises are buying grain at low procurement prices, selling it in the open market, and pocketing the difference.[8] Rather than urban consumers, these bodies could well be the main beneficiaries of the government's costly grain policies.

Advocates of the current system argue that large grain reserves are needed to keep prices stable. But the reserves have not been especially effective at stabilizing prices: witness the rise in food prices in 1993–94.[9] Such increases occur because grain bureaus have little incentive to increase efficiency and release grain promptly when prices rise.[10] In any event, if the purpose is to keep prices stable, targets of 40 million tons in reserves and 90 million tons in total stocks with local and provincial authorities appear excessive.

In most developing countries private traders help stabilize prices automatically by using price signals to sell and buy grain.[11] In addition, importing grain can often be cheaper than buying and storing it. Between 1990 and 1996, for example, China could have saved $35 for each ton of grain it procured and stored by importing it instead.

These considerations suggest two important long-term reforms of China's grain policies. First, the need to inject competition into procurement and distribution, similar in style and purpose to the government's support for nonstate firms entering industry in the 1980s. The government could aim at gradually reducing its grain purchases from today's 75 percent of the marketed total to 25 percent by 2020, leaving room for nonstate trading companies to expand purchasing and retail networks.

Second, the government could steadily remove barriers to the import and export of grain. Doing so would not lead to an unhealthy dependence on imports (see the section below on international trade), and it would give China access to foodgrains at prices significantly below its own marginal costs of production (box 5.1).

Reforming grain policies would also mean dismantling the quota system for grain production. This move would enable farmers to use some of their land for crops that offer bigger returns, accelerating a process

FIGURE 5.2

Procurement prices have been significantly below market prices

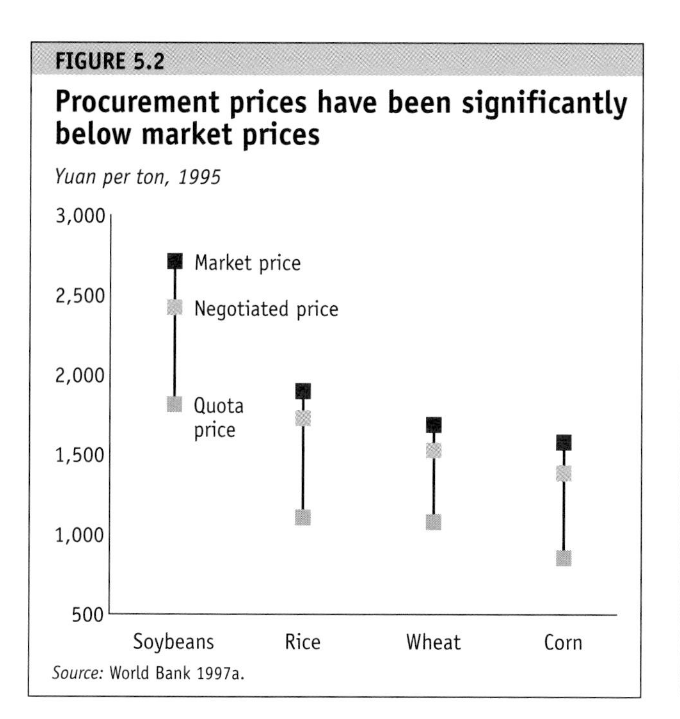

Source: World Bank 1997a.

BOX 5.1

Self-sufficiency will require protection

Grain procurement prices in China are now close to international levels and temporarily above market prices, and the government plans to procure 1997 quotas at market prices. But using further price incentives to stimulate agricultural production would mean raising domestic prices *above* international levels, thereby subsidizing grain producers. According to one simulation, maintaining the government's target of 95 percent grain self-sufficiency would imply that by 2020 the domestic price of grain would need to be 40 percent higher than the world price.

that has already begun (figure 5.3). Since many of China's poor tend to be grain farmers, the opportunity to diversify would lead to significant welfare gains.

Reforming the market for fertilizer

Fertilizer use in China has quadrupled since 1978, spurred by plantings of new crops that respond well to fertilizers. Yet the average application rate—155 kilograms per hectare—is below the average for East Asian developing countries and far below rates in Japan and Korea, where yields are higher.

FIGURE 5.3

Going against the grain

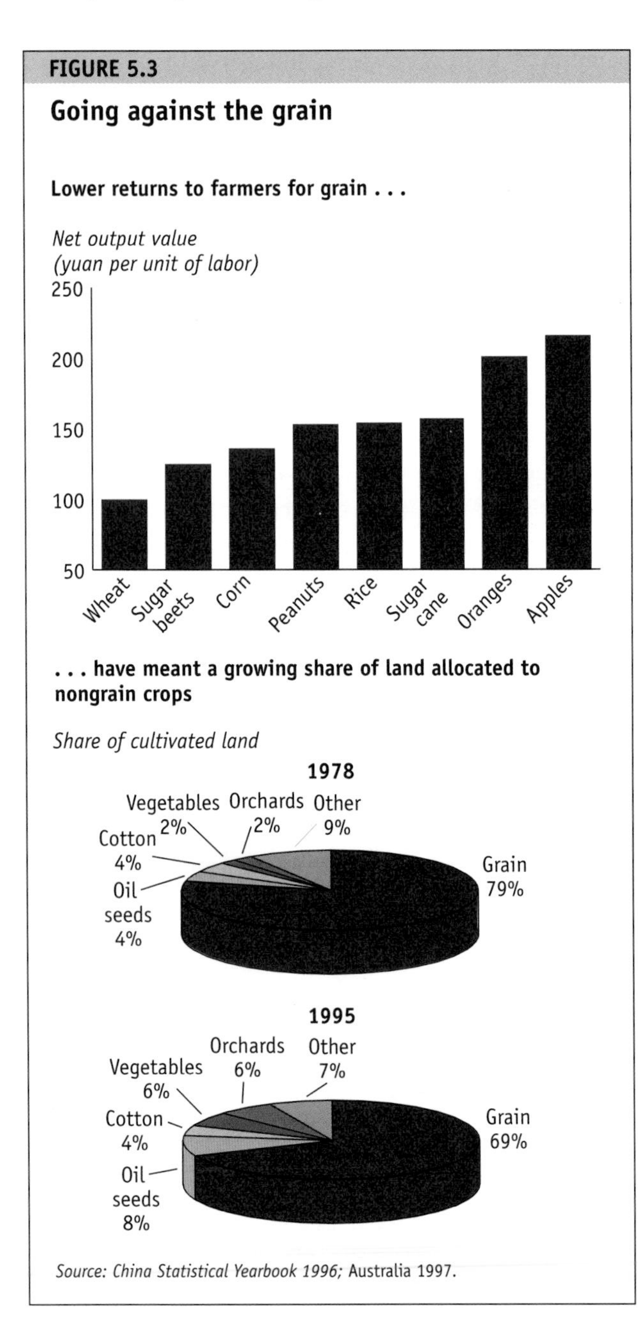

Source: China Statistical Yearbook 1996; Australia 1997.

At the same time, the marginal effect of fertilizer on yields has been declining in China. This is not happening because the potential for yield increases is limited, but because farmers overuse nitrogen and phosphatic fertilizer and underuse potash (which helps plants absorb nitrogen and phosphate). A more balanced use of these fertilizers could increase yields by another 12–15 percent.

Why is potash underused in China, especially since the benefits of a balanced mix of fertilizers are well known? One possible explanation could be that whereas nitrogen and phosphates are produced locally, potash has to be imported. Since fertilizer imports are controlled by the state monopoly Sinochem (under the Ministry of Foreign Trade and Economic Cooperation), imports of potash are restricted.

China's production of nitrogen and phosphate fertilizer using local technologies has been a remarkable feat. But some fertilizer factories are inefficient and polluting, and produce low-grade fertilizer. For example, the domestically produced nitrogen fertilizer, ammonium bicarbonate, can be a serious pollutant and tends to decompose in storage. When it is applied to irrigated fields, about half the fertilizer evaporates. The phosphate fertilizer, a single superphosphate, is produced from low-grade rock and has a low nutrient content.

It will take time for China to develop a modern fertilizer industry. However, it already has sufficient feedstock to manufacture large amounts of urea, a nitrogen fertilizer. It now needs to offer incentives for international joint ventures in raw material mining and develop all the downstream elements of a modern fertilizer industry.

Improving agricultural research and extension

Research and extension have done much to boost China's agricultural productivity. The internal rate of return to investment in agricultural research is estimated at 94 percent.[12] Yet spending on agricultural research relative to agricultural output has declined steadily over the past fifteen years.[13] Moreover, new technologies require increasingly expensive research (genetics and biotechnology being two good examples)—yet budgetary funds per active research scientist have dwindled (figure 5.4).

This budgetary squeeze has encouraged research institutes to seek income from commercial activities.

Ideally, the institutes would be licensing the technology they develop, but they are deterred from doing so by weak protection for intellectual property rights. So they have instead turned to business undertakings, including manufacturing, restaurants, hotels, and trade. Only about 15 percent of net revenues from these activities is allocated to research. Although there is insufficient evidence to conclude that research productivity is declining, it would be surprising if it were not.

Nor do the problems end there. Researchers have great difficulty transferring technology from lab to land because of a shortage of trained and motivated extension workers. Researchers estimate that only a third of their research results ever reach farmers' fields; the rest stay "on the shelf."

These serious shortcomings must be tackled by big increases in budgetary funding for agricultural research and extension, and partly by better use of available funds. For example, the growing demand for feedgrains suggests that research should be switched to developing corn varieties that are suited to rice-growing areas. In addition, a research system organized along agro-ecological zones (rather than administrative boundaries) would make for better use of resources.

Expanding irrigation

Much more than land, water is the biggest constraint on China's agriculture. Water runoff is below the world average, only about a third is exploitable, and most of this is geographically concentrated. The area south of the Yangtze River has 7.5 times more water per square kilometer than the area north of the river (figure 5.5). In the south 450 million people—a third of the population—live under threat of flood; in the north 300 counties and 479 cities are short of water.

Partly because of this uneven distribution, only 60 percent of exploitable water is actually used. Increasing this share is important not just for food security but also for domestic and industrial use. In agriculture alone the shortage of water is estimated to be about 30 billion cubic meters (equal to total use in the Philippines and to about 5 percent of China's total available water). The shortage will double if China increases irrigated farmland area as planned. Groundwater is being tapped to make up the difference, and evidence abounds of overexploitation (subsidence, seawater intrusion, drying up of shallow aquifers, falling water tables).

Despite the shortage of water, its use continues to be wasteful (figure 5.6). Most of the irrigation and drainage schemes constructed in the 1950s and 1960s were hastily designed, poorly built, and often left unfinished. Today these systems are badly run and maintained because of fragmented responsibility among levels of government, no direct participation by farmers in decisionmaking, inadequate budgets, and water charges that are too low to cover maintenance costs.

More public investment and new pricing policies are essential. The investments with the highest return include constructing flood warning and management systems, rehabilitating dams to improve safety, and strengthening containment dikes. These investments would save lives and property and reduce the loss of

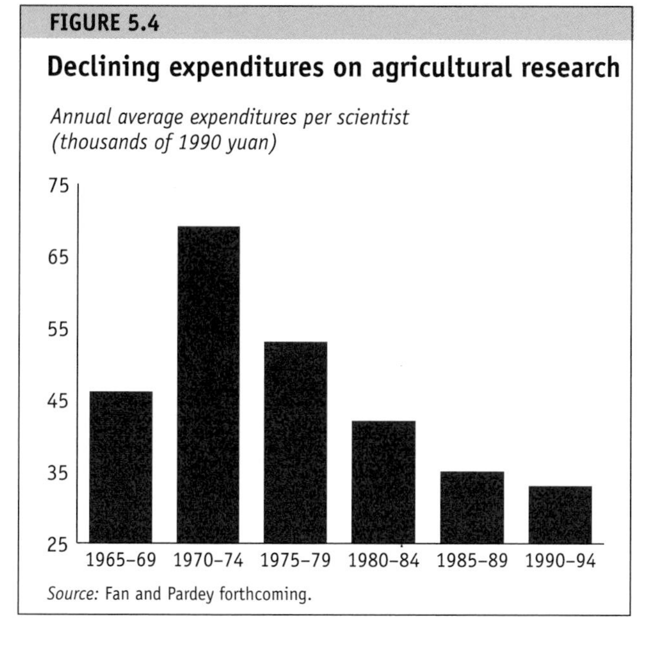

Source: Fan and Pardey forthcoming.

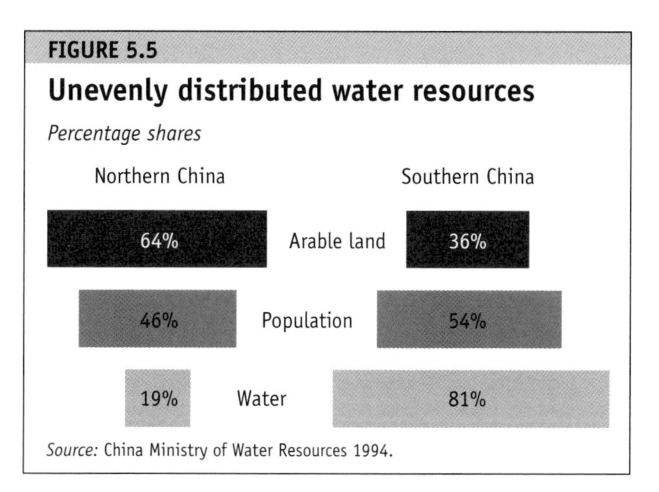

Source: China Ministry of Water Resources 1994.

FIGURE 5.6

Where does the water go?

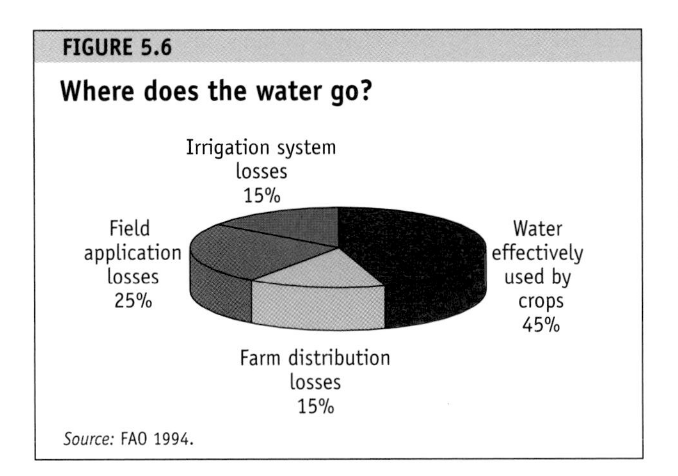

Source: FAO 1994.

BOX 5.2

Another Grand Canal

Managing the course of China's waterways has been a preoccupation for thousands of years, as successive dynasties have sought to control floodwaters, provide irrigation, and facilitate transportation along inland waterways. The centerpiece of these historical engineering achievements is the Grand Canal, which stretches 1,794 kilometers from Hangzhou to Beijing and was completed in the seventh century under the Sui dynasty. Since the founding of the People's Republic of China, the pace of investment in water resource development projects has accelerated—more major dams were constructed in China during this period than in the rest of the world combined.*

In 1995 the Chinese government authorized work on the middle route of the South-North Water Transfer Project, designed to alleviate water shortages in China's north by diverting the waters of the Yangtze River Basin toward Beijing and Tianjin. This project will construct a new "Grand Canal" 1,241 kilometers long, extending from Danjiangkou Reservoir in central China to Beijing, and require the relocation of 200,000 people. While completing this project will be a significant engineering feat, it will not provide a long-term solution to northern China's water scarcity unless the incremental water is appropriately allocated and priced in a way that reflects its long-run marginal and social cost.

* LeMoigne 1990.

crops to floods. High social returns could also be obtained from rehabilitating older irrigation systems. Indeed, if user charges were raised at the same time, these investments could yield handsome financial returns as well. And much could be gained by relining canals and using low-pressure pipelines on existing systems.

To alleviate water shortages in the north, the south-north transfer scheme must rank as a public investment priority (box 5.2).[14] Prefeasibility studies indicate that transferring water along the scheme's eastern and middle routes will cost about 0.4 yuan per cubic meter of water. With estimated average returns to water use of 67 yuan per cubic meter in industry and 2–4 yuan in agriculture, the project's viability is abundantly clear (table 5.1).[15]

A crucial complement to increased investment is the pricing of water at its long-run marginal cost. Higher charges will encourage conservation and raise the necessary resources for public investments. In the few places where the price of irrigation water has been raised, the demand for water has declined without affecting yields. And proper pricing will help produce a more efficient spread of industry and agriculture across the country. Other things being equal, and over the long term, water-intensive agriculture and enterprises will tend to develop south of the Yangtze Basin and less water-intensive activity will shift to the north.

Reclaiming and developing land

Although land is not the main constraint on agriculture, the amount available for cultivation has been reduced by two trends. The first is environmental degradation in all its forms. For example, some 3.7 million square kilometers of land—an area larger than Western Europe—suffer from water and wind erosion. Each year about 0.5 centimeter of topsoil is eroded from the 13 million hectares of mountains and rolling hills in the North China Plain, where some 250 million people live.

The second reason for declining cultivable acreage is construction—of houses, factories, and roads. Since 1988, 190,000 hectares of farmland have been taken over each year. At the same time, about 245,000 hectares of land a year have been reclaimed for farming, but it is of poorer quality and therefore less productive.

There is still considerable scope for reclamation: perhaps a total of 18 million hectares of barren, tidal, and waste lands. Since reclamation will cost between 15,000 and 150,000 yuan a hectare, the government could save itself some money by boosting private involvement through extended leases or outright ownership, and parallel development of land for nonagricultural uses.

TABLE 5.1
High returns to water transfer
(yuan per cubic meter)

Sector	Marginal return to water use
Agriculture	3.1
Electricity	25.7
Chemicals	26.4
Metal	66.2
Other industry	82.3
Coal	163.4
Memo item	
Cost of water transfer	0.4

Source: World Bank 1997b.

Such measures would encourage more private investment in agriculture in general. But it will take time to develop a legal framework and institutions that underpin the rights and responsibilities of tenants and leaseholders on the one hand and lessors and renters on the other. Meanwhile, the government could consider extending the minimum term of tenure from fifteen to thirty years and ensure that this extension is enforced and respected by village authorities.

Projecting grain balances

The future of foodgrain production in China will depend on the government's progress on the five challenges described above. In scenarios explored by the World Bank, the difference between success (the high case) and business-as-usual (the low case) could be as much as 60 million tons in 2020, or 13–15 percent of today's grain consumption (table 5.2).[16]

The most likely scenario—labeled the base case in table 5.2—shows China being able to produce 636 million tons of grain in 2020.[17] This scenario assumes reasonably liberal trade policies consistent with entry into the World Trade Organization, completion of two south-north water transfer routes, balanced fertilizer application, and adequate investments in water and research.

Projecting demand is equally (perhaps more) difficult. China's enormous population means that small changes in per capita consumption, income growth, or demand elasticities for foodgrains and meat can result in large changes in total demand. Although the estimates in the three scenarios for 2020 are based on plausible assumptions of these parameters, they should be

TABLE 5.2
Alternative scenarios for future grain supply
(millions of tons of unmilled grain)

		2020 (projected)		
Measure	1996	High case	Base case	Low case
Production	416	667	636	606
Rice	185	313	298	283
Wheat	102	151	144	137
Coarse grain	129	203	194	186
Consumption	437	695	695	695
Rice	187	298	298	298
Wheat	114	162	162	162
Coarse grain	136	235	235	235
Imports	−21	−28	−59	−89
Rice	−2	14	0	−16
Wheat	−12	−11	−18	−25
Coarse grain	−7	−32	−41	−49

Source: World Bank 1997a; *China Statistical Yearbook 1996.*

viewed with considerable caution. Caution is even more necessary for estimates of imports, which are the small difference between two large numbers: if base case demand is underestimated by 5 percent and production overestimated by 5 percent, imports would be 212 percent larger than is projected here.

The growth rates projected for production and consumption under the base case scenario are near the top of the range of estimates made by other researchers (table 5.3), though end-year imports are broadly similar.

Using international trade

A key message of these projections is that China will need to deepen its engagement with world grain markets if it wants its agriculture to be efficient. The alternative—self-sufficiency behind trade barriers—would lock the economy into high subsidies for agriculture and jeopardize rapid, sustainable growth. As Japan and Europe have learned, subsidies are difficult to withdraw and can inhibit structural change and adjustment.

Dealing with world grain markets

The government is concerned that increasing grain imports could make China vulnerable to volatile world prices and unreliable suppliers. While these concerns are valid, they can be diminished and should be weighed against the risks associated with relying solely on domestic production.

TABLE 5.3
Comparing projections
(percentage average annual growth)

	Projection period	Growth in grain production	Growth in grain demand	End-year imports (millions of tons)
World Bank 1997a	1992–2020	1.8–2.2	2.6	28–89
Rosegrant and others 1995	1990–2020	1.5	1.5	22.0
Brown 1995	1990–2030	–0.6	0.8–1.6	207–369
USDA/ERS 1994	1994–2006	1.1	1.2	20.0
Garnaut and Ma 1992	1990–2000	1.1	1.9–2.7	50–90
World Bank 1993	1990–2000	1.6	2.4	11.3
	2000–2010	1.6	1.8	21.6

China currently imports about 5 percent of the grain it needs. By 2020 grain imports could be as low as 4 percent and as high as 14 percent. Since world trade in grain is only about 200 million tons a year and unlikely to grow rapidly, China could by then be a large presence in the international market.

But this change would occur gradually, so foreign suppliers would have time to adjust. The main grain exporters have been highly responsive to global demand, partly because their excess capacity is growing.[18] Stagnant world demand over the past two decades has pulled down prices, especially for wheat and rice (figure 5.7), so Western exporters have withdrawn an estimated 34.5 million hectares from grain production. With the right incentives this idle capacity (of 115 million tons of grain) could be returned to production quickly. The same is probably true of the former Soviet Union (where production declined by 86 million tons between 1990 and 1995) and Argentina.

If China were to increase its imports of grain gradually and consistently, it could develop long-term contractual relationships with different suppliers. This has never been tried before, so China and its trading partners would need to be pioneers, perhaps under the aegis of the World Trade Organization. Both sides would certainly benefit—China from increased certainty of supplies, and exporters from reduced risks.

Even so, China would still have to rely on spot markets for some of its grain imports. Futures and forward markets for grain are so thin for longer-dated maturities that they provide little comfort for a large importer. Thus China will have to absorb the cost of this volatility, although with a flexible system of private trading and adequate strategic reserves, domestic prices are unlikely to be any more volatile than they are already.

Improving trade infrastructure

Although more imports are unlikely to make China vulnerable to international markets, they will strain its infrastructure. About 98 percent of the grain trade flows through fourteen ports, and only three are specialized in grain handling. Bulk loading facilities are generally lacking, and there are no intermodal bulk transfer systems, except between Beijing and Tianjin. Off-loading is slow, which raises costs by about $15 a ton; were China to import 60 million tons by 2020, this extra cost alone would be $900 million a year at today's prices. Over the next twenty years China will need ten new deepwater bulk grain berths, each with a capacity of 4–5 million tons. It will also need more high-volume rail corridors using bulk rail wagons and bulk loading and unloading facilities at railheads. Similarly, transit

FIGURE 5.7
Declining world prices for wheat and rice

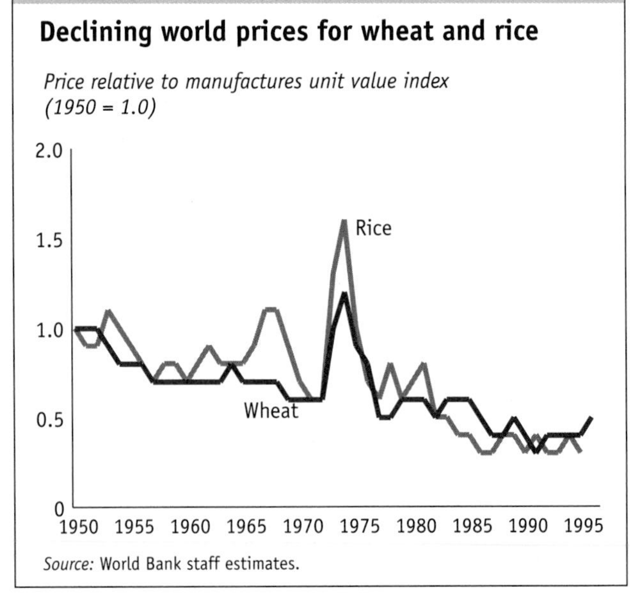

Source: World Bank staff estimates.

storage will have to cope with high throughput not just of imported grain, but also of domestic grain for domestic and foreign markets.

Notes

1. This chapter draws heavily on World Bank (1997a).

2. Brown (1995).

3. Rosegrant, Agcaoili-Sombilla, and Perez (1995).

4. For consistency in consumption, production, and import data, and to remain consistent with Chinese production data, weight measures for rice are actually for paddy (which is roughly 43 percent heavier).

5. This strategy is explained in a government White Paper, "The Grain Issue in China" (October 1996), available from the Information Office of the State Council.

6. Annual consumption subsidies for grain and edible oils averaged 20 billion yuan during 1992–96, while annual debts of grain enterprises grew by 25 billion yuan in 1996.

7. The implicit tax imposed by below-procurement prices equals 40 billion yuan. The total subsidy to grain enterprises equals 45 billion yuan (see note 6). The total comes to 85 billion yuan. This 85 billion yuan is equal to 6 percent of total rural per capita income in 1995 (860 million rural inhabitants with an average per capita income of 1,578 yuan). Annual 8–7 Plan spending increased from about 10 billion yuan in 1994–96 to 15 billion yuan in 1997. So, for 1995 the 85 billion yuan was eight times 8–7 Plan spending. Beginning in 1997 it was almost six times that amount. (The objective of the 8–7 Plan, announced in 1993, is to reduce the number of absolute poor in China from the 1993 level of 80 million to zero over the 1994–2000 period).

8. Australia (1997).

9. Analysis suggests that the sudden rise in food prices during this period had little to do with the decline in grain output.

10. In fact, grain enterprises are paid for *storing* grain, and so are reluctant to release it when they are required to do so.

11. Indonesia, for example, relies heavily (though not exclusively) on private traders. BULOG, the government's procurement agency, has only a modest market role, purchasing less than a quarter of rice production and holding carryover stocks of about 1 million tons.

12. Fan (1996).

13. It is worth noting that while this ratio is higher than in most developing countries, it is a quarter of the ratio in industrial countries.

14. See *Chinese Environment and Development*, summer 1994, for a review of technical and environmental issues in the South-North Water Transfer Project. See the *People's Daily*, January 7, 1995, for an announcement of the approval of the project.

15. World Bank (1997b).

16. For a detailed discussion of these projections and the analytical model used to arrive at these conclusions, see World Bank 1997a.

17. This estimate uses the unmilled equivalent (paddy) for rice.

18. The major grain exporters are the United States (42 percent of the world market), European Union (22 percent), Canada (11 percent), Australia (7 percent), Argentina (6 percent), and Thailand (2 percent).

References

Australia, Department of Foreign Affairs and Trade, East Asia Analytical Unit. 1997. *China Embraces the Market: Achievements, Constraints, and Opportunities.* Canberra.

Brown, L.R. 1995. *Who Will Feed China? Wake-Up Call for a Small Planet.* New York: W.W. Norton.

China Ministry of Water Resources. 1994. "China's Water Resources." Beijing.

Fan, S. 1996. "Research Investment, Input Quality, and the Economic Returns to Chinese Agriculture." Paper presented at the post-conference workshop on Agricultural Productivity and Research and Development Policy in China, August 29, Melbourne, Australia.

Fan, S., and P. Pardey. Forthcoming. "Research Productivity and Output Growth in Chinese Agriculture." *Journal of Development Economics.*

FAO (Food and Agriculture Organization). 1994. *Water Is Life.* Rome.

Garnaut, R., and G. Ma. 1992. *Grain in China: A Report.* Australia Department of Foreign Affairs and Trade, East Asian Analytical Unit, Canberra, Australia.

LeMoigne, Guy. 1990. "Keynote Address." Made to Senior Policy Seminar on Policies for Multipurpose River Basin Development in China, March 31–April 21, Nanjing, China.

Rosegrant, M.W., M. Agcaoili-Sombilla, and N.D. Perez. 1995. *Global Food Implications to 2020: Implications for Investment.* Food, Agriculture, and the Environment Discussion Paper 5. Washington D.C.: International Food Policy Research Institute.

USDA/ERS (U.S. Department of Agriculture, Economic Research Service). 1994. Cited in J. Huang, S. Rozelle, and M.W. Rosegrant, eds., *Supply, Demand, and China's Future Grain Deficit.* Food, Agriculture and the Environment Discussion Paper. Washington D.C.: International Food Policy Research Institute.

World Bank. 1993. "The World Food Outlook." International Economics Department, Washington D.C.

———. 1997a. *At China's Table: Food Security Options* Washington, D.C.

———. 1997b. "China: Wanjiazhai Water Transfer Project." Staff Appraisal Report 15999-CHA. China and Mongolia Department, Washington, D.C.

chapter six

Protecting the Environment

China's growth has brought dramatic improvements in living standards and serious damage to its environment. But future growth need not be purchased at the cost of higher pollution. With policies that harness markets, encourage alternatives, and provide an appropriate regulatory framework, China can grow both richer and cleaner.[1]

China's air and water, particularly in urban areas, are among the most polluted in the world. Ambient concentrations of most pollutants exceed international standards several times over, burdening China with vast human and economic costs. As many as 289,000 deaths a year could be avoided if air pollution alone were reduced to comply with Chinese government standards. Overall, the economic costs of China's air and water pollution have been estimated at 3–8 percent of GDP a year.

Two forces are responsible for much of China's environmental degradation and will remain crucial well into the next century. The first is China's extreme dependence on coal. Today coal satisfies nearly 80 percent of China's burgeoning demand for energy, making China the world's largest coal consumer. Abundant coal reserves, combined with an understandable reluctance to rely on imports of cleaner oil and natural gas, ensure that coal will be China's main source of energy for years to come. The second factor is China's booming cities. Between 1978 and 1995 their population swelled by 180 million residents, plus some 50 million unregistered migrants from the countryside. Rising urbanization has not only been accompanied by increased automobile use and largely untreated emissions of municipal waste, it has also increased the portion of the population exposed to the greater pollution found in urban areas.

The government recognizes the environmental challenges confronting the country and over the past decade has introduced a comprehensive legal framework to protect the environment.[2] These efforts can claim some successes: for example, pollution intensities—emissions per unit of output—have fallen in recent years.[3] Still, much remains to be done. In the words of Premier Li Peng, "We clearly are aware that the situation of the environment in our country is still quite severe. . . . Environmental pollution in cities is worsening and extending into rural areas, and the scope of ecological damage is increasing."[4] This chapter lays out China's environmental challenges and discusses policy alternatives for the next twenty-five years.

Deteriorating environmental conditions

Over the past two decades urbanization, industrialization, and motorization have seriously damaged air and water quality. Moreover, increasingly intensive agricultural practices have produced a new generation of environmental threats. Runoff from fertilized fields has contributed to water pollution, and heavy and often inefficient irrigation has exacerbated water shortages and salinized large tracts of land. Efforts to bring marginal lands under cultivation have worsened soil erosion and desertification and threatened China's fragile wetlands and grasslands. In the face of these many challenges, setting priorities for a cleaner future is difficult. But, particularly for air and water pollution, the benefits of abatement so clearly outweigh the costs that they justify action under any circumstances.

Air pollution

Although particulate emissions have remained roughly constant since the 1980s (no mean feat, considering that coal consumption has more than doubled), sulfur dioxide emissions have soared. Ambient levels of particulates and sulfur dioxide are now among the highest in the world (figure 6.1).[5] Although systematic data are scarce, ambient lead levels also appear to be rising: recent evidence suggests that as many as half the children in parts of Shanghai suffer from elevated levels of lead in their blood.

Three culprits are responsible for the bulk of China's urban air pollution. First, small and relatively inefficient coal-fired industrial boilers, which often vent their emissions from low stacks, account for between one-third and one-half of ground-level particulates and sulfur dioxide. Second is residential use of coal. Although residential use accounts for just 15 percent of total coal use, it is responsible for a further one-third of particulates and sulfur dioxides. It also affects the quality of indoor air, causing a health hazard roughly as serious as smoking. Third, in major cities the number of cars is

FIGURE 6.1

Air pollution in China's cities is high, even by developing country standards

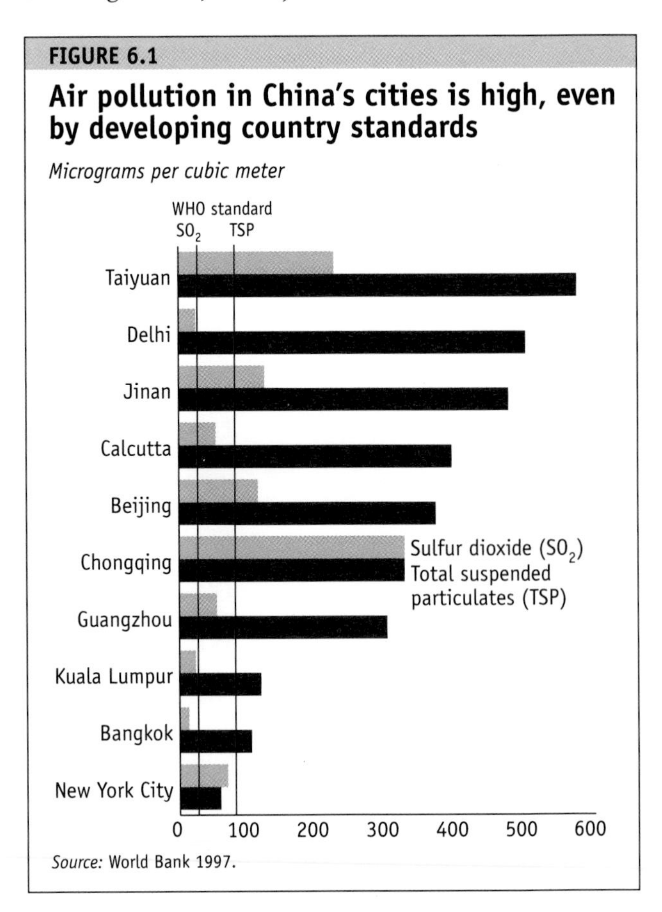

Source: World Bank 1997.

growing by 10 percent a year. This contributes the balance of particulate and sulfur dioxide emissions, and is also a rising source of lead in the urban atmosphere.[6] The damage is aggravated by slow average driving speeds and low vehicular emissions standards. As a result, although Beijing has only one-tenth the number of automobiles of Los Angeles, its automotive emissions are almost as great.

All this pollution is bad for health. Mortality rates from chronic obstructive pulmonary disease, the leading cause of death in China, are five times those in the United States. This is not simply due to China's much higher rates of smoking and poorer health care. Epidemiological studies indicate that the difference in particulate and sulfur dioxide concentrations between Beijing and New York City (about 300 and 50 micrograms per cubic meter, respectively) is associated with 130 percent higher mortality rates from chronic obstructive pulmonary disease.[7] Reducing outdoor air pollution to target levels set by the Chinese government would save 178,000 lives a year. Doing the same for indoor air pollution would save another 110,000.

But the costs of air pollution go well beyond this tragic—and needless—loss of life. They also include an estimated 566,000 additional hospital admissions and nearly 11 million emergency room visits directly attributable to air pollution that exceeds China's standards. This extra burden strains China's health care system and results in lost working time equivalent to 7.4 million person-years each year.[8] Numerous studies in China and elsewhere have also documented the damage to children of lead exposure, from stunted growth to neurobiological disorders and intelligence quotient (IQ) deficiencies.

Acid rain is another costly byproduct of China's air pollution. Emissions of sulfur dioxide and nitrous oxide react with atmospheric water and oxygen to form sulfuric acid and nitric acid, which can return to earth nearby or even thousands of miles away from the source of air pollution. Acid rain causes crop damage, deforestation, structural damage to buildings, and harm to human health. These consequences have been particularly severe in southern China, where high-sulfur coal is burned in large quantities. A study by the Chongqing Environmental Protection Bureau found that nearly one-quarter of the vegetable crops in the Chongqing area were damaged by acid rain in 1993.

Water pollution

Industrial and municipal waste and chemical and organic fertilizer runoff are the main sources of water pollution in China. Reported discharges of effluent have increased moderately since the early 1980s, with municipal waste accounting for a rising share of total discharges (figure 6.2). But the total increase is probably much larger, since sources in the rapidly growing nonstate sector are monitored only sporadically. The rising share of municipal waste is of particular concern, since less than 20 percent receives any form of treatment (compared with nearly three-quarters of industrial waste).

Although water pollution is linked with many diseases, its health impact in China has generally been contained by widespread access to safe drinking water and sanitation. Common water-related diseases account for only a small fraction of total morbidity and disease.[9] Still, increasing pollution has raised the costs of providing drinking water. For example, pollution caused by untreated discharges into the Huangpu River has forced the municipality of Shanghai to move its drinking water sources upstream, at a cost of $300 million. In many cities households must boil their water before drinking it, which is far more costly than centralized chlorination.

The more serious concern about water pollution is that it aggravates China's water shortage (see chapter 5). Water pollution exacerbates water shortages in cities

FIGURE 6.2

Cities account for a rising share of wastewater discharge

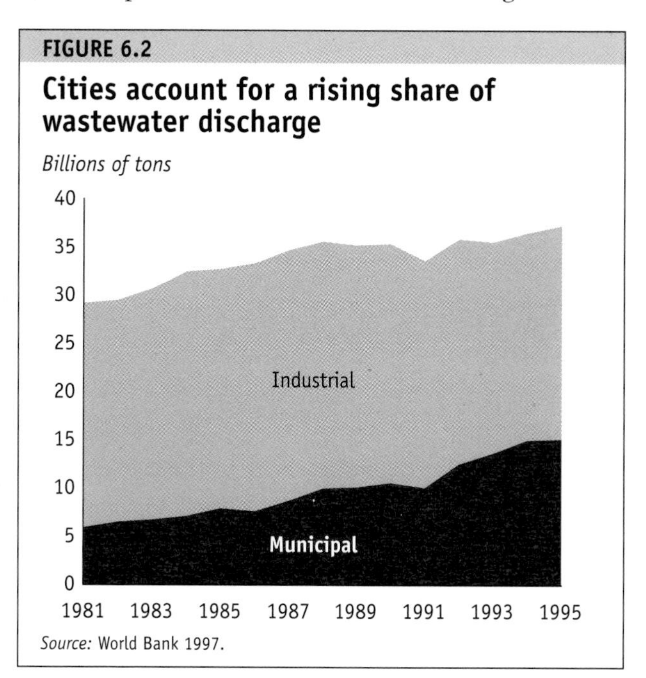

Source: World Bank 1997.

and towns, and increased water supply and consumption tend to result in greater water pollution unless effluent treatment is provided. In 1993 about 8 percent of agricultural lands received water so polluted that it was unfit for use, leading to an estimated loss in grain production of 1 million tons. If wastewater treatment were improved from the current 20 percent to 50 percent, total grain production could increase by 24 million tons by 2020.[10]

Pollution prospects

China is one of the most polluted countries in the world. If past trends continue, its environmental future will be bleak. But China can also grow cleaner. Policies that encourage improvements in energy efficiency and energy conservation, combined with substitution away from coal, widespread adoption of technologies for combating air and water pollution, and reliance on public transit rather than automobiles, would considerably brighten China's environmental prospects.

Two aspects of continued economic development will particularly affect China's environmental prospects. The first is energy demand. Although energy intensity—energy consumption per unit of GDP—will probably fall from its current high levels, total energy consumption will rise sharply over the next twenty-five years. In the absence of affordable alternatives, much of this demand will be met by coal (box 6.1).[11] Since coal is the main source of urban air pollution, this presents a major environmental challenge.

Second, if other countries' experiences and China's past are any guide, rising incomes will be accompanied by increased urbanization and rapid growth in automobile use (figure 6.3). Since automobiles contribute the bulk of ambient lead concentrations and are a rising source of other forms of air pollution, this too will shape environmental prospects over the next twenty-five years.

Against this background, consider two alternative scenarios:

- The first scenario assumes no significant changes to current environmental policies or their enforcement. This is by no means the most plausible scenario. China's past progress in designing and implementing environmental policy reforms makes a scenario with no further improvements unlikely. But it does highlight the enormous costs that would ensue were the momentum of environmental policy reforms to falter. Under this "faltering reforms" scenario both the energy intensity and emissions intensity of GDP will fall. Even without new incentives, rapid growth will ensure that outdated and pollution-intensive technologies are replaced swiftly. By 2020 less than 10 percent of China's current capital stock will still be in service. Although emission intensities decline, however, overall pollution will rise sharply (figure 6.4). As a result particulate emissions will increase by 30 percent and sulfur dioxide emissions by 60 percent, with serious consequences for air quality and acid rain. The number of deaths attributable to outdoor urban air pollution alone will rise dramatically, from 180,000 a year now to as many as 600,000 in 2020. However, improvements in the quality of the capital stock should reduce emissions of major water pollutants in absolute terms.

FIGURE 6.3

Driving into the future?

Automobiles per 1,000 people (log scale)

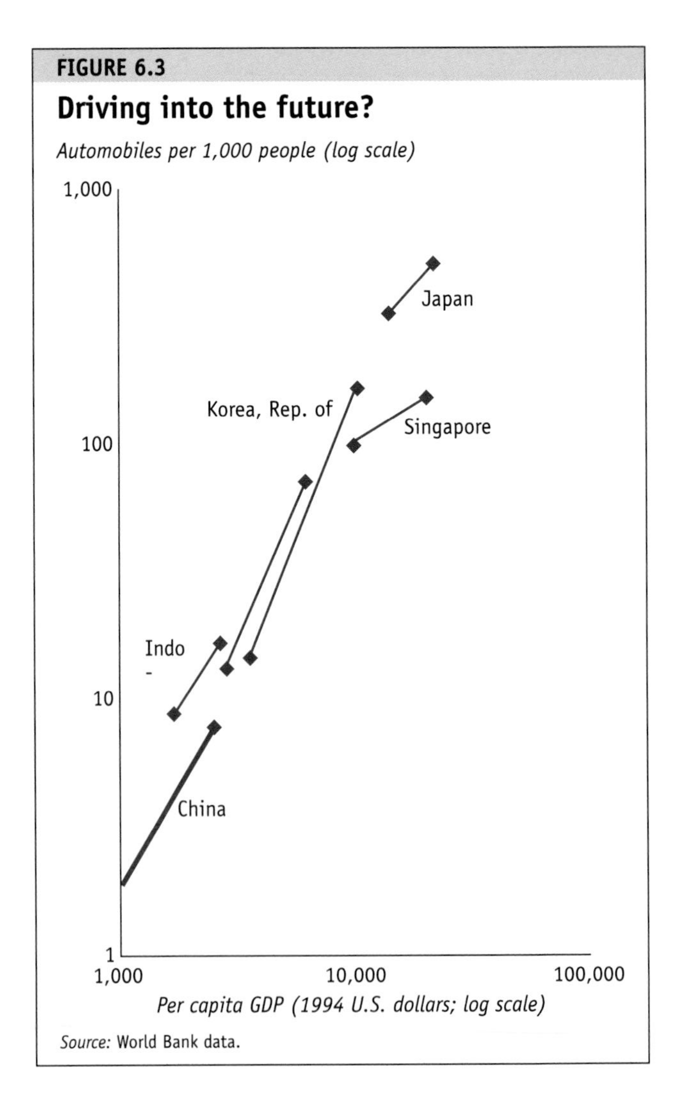

Source: World Bank data.

• The second scenario involves a substantial deepening of reforms. Over time different policies could bring about improvements in energy efficiency and reductions in emissions to roughly U.S. or European standards of the early 1980s. China can grow, and grow cleaner. Under this "deepening reforms" scenario, particulate emissions would be halved, causing mortalities from air pollution to fall to 140,000. These reductions in emissions would be a remarkable achievement for a country as poor as China is today. International experience

BOX 6.1

China: Where coal is—and will be—king

China is the world's largest consumer of coal, burning one in every three tons of coal used worldwide in 1995. Coal accounts for nearly 80 percent of primary commercial energy use and 75 percent of residential use. Strong growth in energy demand over the next twenty-five years, combined with a dearth of affordable alternatives, means that China will continue to burn coal on a massive scale for the foreseeable future.

Despite efficiency gains, energy demand is high and rising. Although China's energy intensity—consumption per unit of GDP—has fallen by half over the past twenty years, it is still between three and ten times that of major industrial economies. But current high energy intensities hold the promise of future declines. Over the next twenty-five years energy intensities are expected to decline to 60 percent of current levels, resulting in energy demand rising by 4.5 percent a year, or about 2 percent a year less than overall GDP growth (top figure). Changes in the structure of GDP, especially at the subsectoral and product levels, will be primarily responsible. With rising incomes, an increasing share of growth will be the result of less energy-intensive improvements in product quality rather than simple increases in the quantity of goods produced.

Few affordable alternatives. At an average annual growth rate of 4.5 percent, total energy consumption will triple over the next twenty-five years. But even a major shift toward alternative sources of energy will not significantly reduce coal dependence. Even if China were to develop every viable hydroelectric power site, hydropower would supply just 8 percent of primary energy demand in 2020. And even if China were to install ten new 600 megawatt nuclear facilities a year, nuclear power would contribute only 6 percent. Although certain renewable energy sources, especially wind power, are particularly promising in China, even major development would meet only a tiny fraction of energy needs. Not only are these alternatives woefully inadequate in the face of energy demand, they are also costly. Even after taking into account the social costs of coal-fired power generation, and ignoring the substantial environmental concerns about massive nuclear and hydroelectric development, China's abundant coal reserves ensure that most alternatives are simply more expensive per unit of power produced (bottom figure).

In fact, the only sizable and affordable alternatives to coal over the medium term are oil and natural gas, primarily from imported sources. China consumes about 150 million tons of oil a year, and its production capacity is estimated to be about 200 million tons a year in 2020. With increased automobile use, oil demand is likely to grow somewhat faster than overall energy demand, so oil imports could easily reach 300 million tons a year in 2020. Although not trivial, the cost of these imports at current prices ($42 billion) will be less than 5 percent of China's expected export earnings of $700 billion.

Source: World Bank 1997.

Robust growth in energy demand

1980 = 1.0

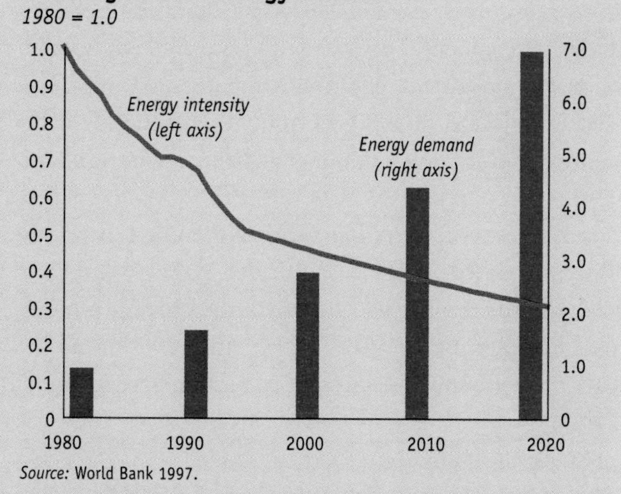

Source: World Bank 1997.

Costly alternatives

Cost per kilowatt hour in 2020
(1990 yuan)

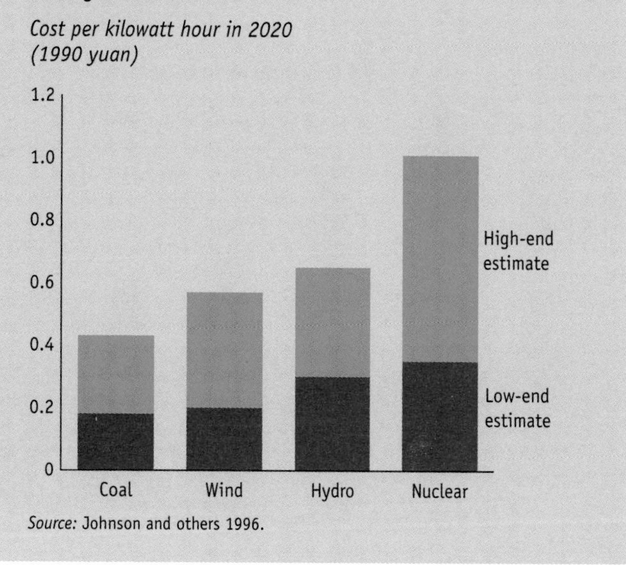

Source: Johnson and others 1996.

shows that pollution initially tends to rise with per capita incomes, and begins to fall only as countries get much richer (figure 6.5).[12] China's per capita income today is well below the "turning points" for most major pollutants. The policies that will help to bring about the unprecedented prospect of a steady decline in pollution beginning immediately are the subject of the rest of this chapter.

Formulating policies for a cleaner future

The magnitude of China's environmental challenges is clear. Moreover, as the widely varying outcomes under alternative scenarios illustrate, the stakes for environmental policy are high. But China is a poor and growing country with many competing claims on public and private resources. Reconciling these claims and prioritizing environmental policies require careful analysis of the economic costs of environmental degradation. This need is reflected in the government's commitment to eventually develop comprehensive environmental national accounts.[13]

In the interim, partial analysis of the costs of pollution can provide a useful guide for policy. Analysis done for this report suggests that the current annual costs of pollution are substantial. Air pollution alone costs the economy 2–6 percent of GDP per year (table 6.1). Water pollution costs a further 1 percent of GDP, and possibly much more, given the difficulties in quantifying the economic damage it causes. Policies that bring about reductions in air and water pollution will yield significant dividends for future generations.

An agenda for action for the next twenty-five years rests on three pillars: harnessing market forces, creating incentives for investment in cleaner production, and developing effective regulation.

Harnessing markets

Market forces have provided the foundations for the economic growth of the past eighteen years. Properly harnessed, they will be crucial allies in the fight for a cleaner future. Deeper reliance on market forces means that prices—modified to reflect social costs—become powerful tools for altering consumer and producer behavior in ways that benefit the environment. Despite recent reforms in natural resource pricing and taxation, however, these tools have not been deployed to full effect. Most prices now reflect production costs, but they are far from reflecting social costs.

Consider coal. Over the past decade coal prices in coastal provinces have risen close to world levels.[14] But administered prices do not yet adequately reflect the wide variation in the sulfur and ash content of domestic coal, despite the much higher pollution costs of high-

FIGURE 6.4

Projected pollutant emissions in 2020

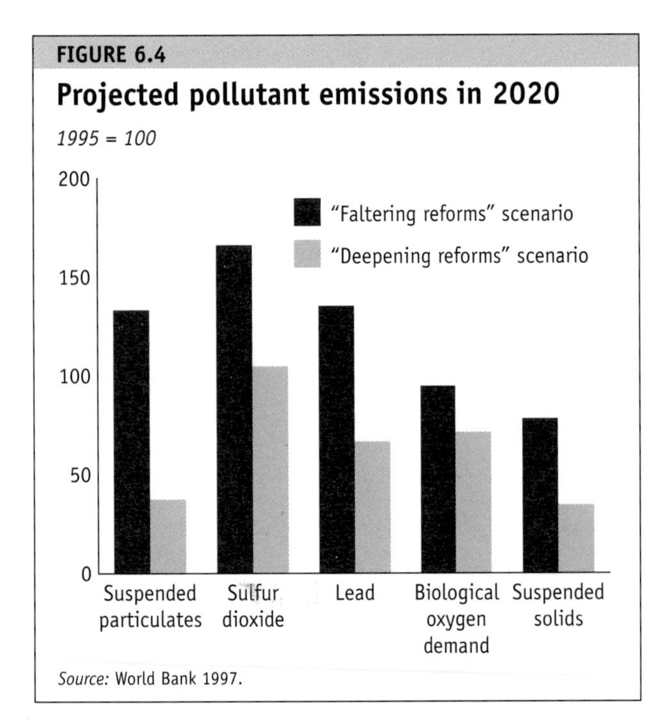

Source: World Bank 1997.

FIGURE 6.5

Across countries, pollution tends to rise, then decline with income

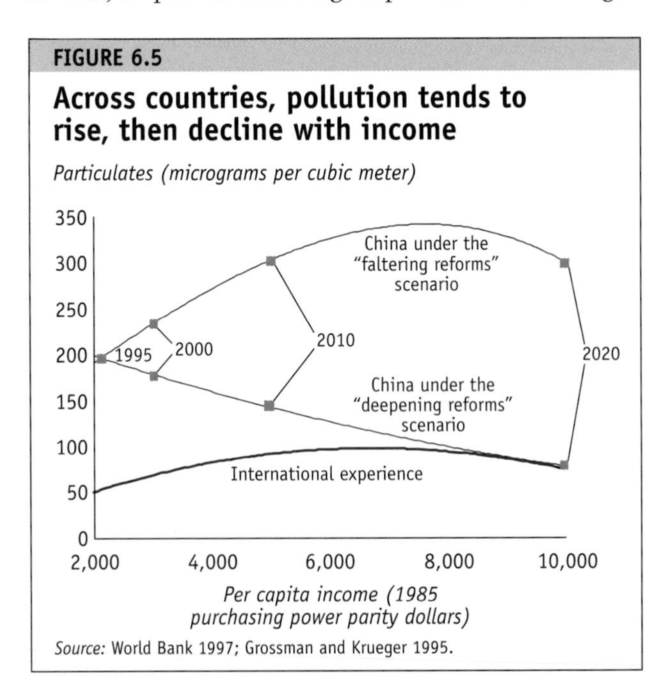

Source: World Bank 1997; Grossman and Krueger 1995.

TABLE 6.1
Costly contaminants: Costs of major forms of pollution, 1995

	Conservative estimate		Moderate estimate	
Type of pollution	Billions of U.S. dollars	Percentage share	Billions of U.S. dollars	Percentage share
Urban air pollution	11.3	48	32.3	61
Indoor air pollution	3.7	16	10.6	20
Lead exposure	0.3	1	1.6	3
Water pollution	3.9	17	3.9	7
Acid rain	4.4	19	4.4	9
Total	24	100	53	100
Percentage of GDP	3.5		7.7	

Note: The conservative and moderate estimates differ in their economic valuation of human life. The conservative estimate uses forgone wages to value human life. The moderate estimate values human life based on conservative estimates of what individuals are willing to pay to avoid risks to life.
Source: World Bank 1997.

sulfur and unwashed coal. Coal prices in interior provinces are generally much lower because of lower transportation costs, which account for as much as 70 percent of the delivered price in coastal areas. But this differential means that interior provinces have little incentive to adopt conservation measures or coal washing technologies that are economical only at higher prices.

As administered prices give way to market-determined prices, the government could consider phasing in coal taxes—based on sulfur and ash content—that more closely reflect social costs. The gradual and preannounced implementation of such a tax would provide incentives to invest in coal-washing capacity and higher-efficiency boilers. Moreover, such a tax would immediately affect the behavior of millions of small industrial and residential users of coal.

Similar arguments apply to gasoline and diesel fuels. Pump prices are low by international standards, and diesel for agricultural use remains heavily subsidized. A liter of gasoline costs $0.87 in the Republic of Korea and $1.13 in Japan, but just $0.28 in China. Narrowing that gap will encourage consumers to economize, and the best way is through phased increases in fuel taxes.

Although price-based incentives can be effective, they are not selective: they do not recognize that the costs of conservation and abatement vary widely across firms. Plant-level data reveal that the marginal cost of reducing particulate emissions is as much as twenty times higher in small firms than in large ones, because of the high fixed costs of abatement technologies (figure 6.6). There are also substantial differences between state and nonstate enterprises, with abatement costs about five times higher in state enterprises.

These contrasts highlight the importance of targeted measures such as emission levies. In addition to providing potent incentives, levies can be tailored to reflect differences in abatement costs across sectors, firms, and regions, and, importantly, the speed of abatement. The National Environmental Protection Agency's recent recommendation of a tenfold increase in the air pollution levy is particularly welcome, given that in several provinces effective air and water pollution levies have actually fallen in real terms since 1987 (figure 6.7). Consistent enforcement of this levy, especially in the nonstate sector (where abatement costs are substantially lower because of low abatement rates among

FIGURE 6.6
Costs of abatement vary widely
Marginal cost of halving particulate emissions

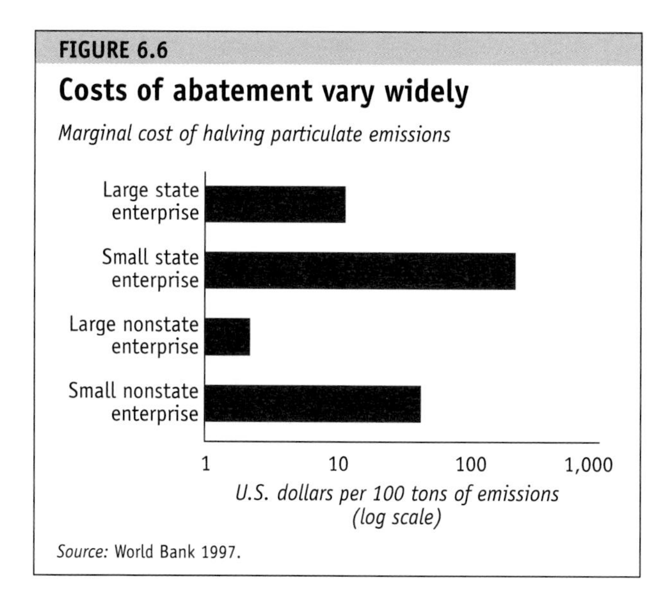

Source: World Bank 1997.

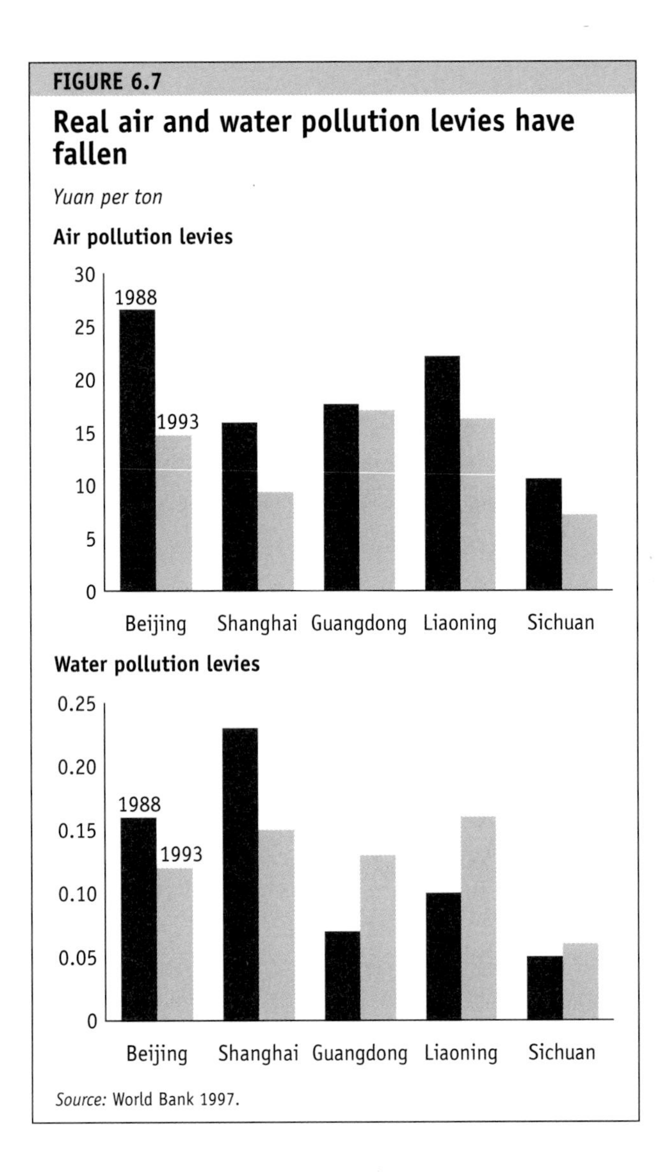

FIGURE 6.7
Real air and water pollution levies have fallen

Source: World Bank 1997.

these firms), would generate large reductions in emissions at relatively low cost.

Consistent application of the principle that the "polluter pays" could have extremely high returns. This is illustrated by Zhengzhou, the capital of Hunan province, where air pollution is typical of China as a whole.[15] Zhengzhou's industries pour 45,000 tons of sulfur dioxide into the air every year, contributing to ambient concentrations of 90 micrograms per cubic meter—about two times the World Health Organization standard. At current low rates of abatement, the incremental cost of reducing sulfur dioxide emissions by 1 ton is only $1.70, and would save 0.63 percent of a "statistical life." This means that pollution levies in Zhengzhou implicitly value a human life at only $270 ($1.70/.00625). Since this is less than even the starkest valuation of human life, the returns to higher pollution levies and better enforcement are very high indeed. In fact, the optimal levy in Zhengzhou—the levy at which the cost to firms of abatement equals the social benefit of cleaner air—is estimated to be nearly fifty times the current level.

The potential effectiveness of price-based incentives depends largely on the government's success in deepening market reforms. Trade liberalization and state enterprise reform are particularly relevant. Removing the remaining obstacles to domestic and international trade would encourage increased specialization, allowing the realization of scale economies in both production and pollution abatement. Liberalizing international trade would also boost access to state-of-the-art capital goods imports with higher standards of energy efficiency and pollutant generation than their domestically produced counterparts. Finally, further state enterprise reforms will help make market-based incentives effective. Without hardened budget constraints, these enterprises will be no more responsive to environment-friendly incentives than to any other price signals.

Investing in alternatives

Investment in alternative energy sources is an essential complement to a pricing system that reflects the social costs of pollution. Without such investment, consumers will be stuck with the worst of both worlds—high prices and continued heavy pollution. Of course, by themselves price-based policies will create strong incentives for investment in a range of energy conservation and emissions abatement technologies with large private and social returns. For example, raising the energy efficiency of industrial energy consumers to OECD standards could reduce coal use by 250 million tons, or one-sixth of current consumption (figure 6.8).

Investments in public infrastructure would also yield large environmental dividends. Examples include boosting natural gas availability for urban residents, public transit, and treatment facilities for industrial and municipal wastewater.

- Boosting the availability of gas for households would dramatically reduce indoor and outdoor particulate and sulfur dioxide concentrations as gas replaced coal for cooking and heating. Moreover, gas is up to five times as efficient as coal and is much preferred by households. Thus gas provision would likely be highly profitable with even moderate increases in taxes on alternatives such as coal. This suggests that there is ample scope for private and foreign participation in the construction of gas infrastructure for urban distribution.

- Better public transport would do much to improve the quality of urban life. International experience shows that China cannot build its way out of congestion through more roads. Rather, transport investments should focus on establishing fixed-link services in high-traffic corridors where high demand will ensure cost-effectiveness, and investing in integrated bus systems that provide an attractive alternative to automobiles.
- Investing in water supply and treatment will pay large dividends. Dalian's experience is illustrative. Nearly half of Dalian's municipal water supplies are consumed by industry, and several large factories used to close for up to two months during the dry season. In 1992 a new wastewater treatment facility enabled the city to provide treated water at one-third the cost of new industrial tap water. It also reduced emissions of wastewater, and factories no longer had to close. Over the long term the profit implicit in these projects should stimulate heavy private and foreign participation. The tendering in early 1997 of China's first municipal build-operate-transfer (BOT) water project, the Chengdu No. 6 Water Plant, is a step in this direction.

FIGURE 6.8

Much room for improvement in energy efficiency

Potential energy savings if efficiency were raised to OECD standards (OECD = 100)

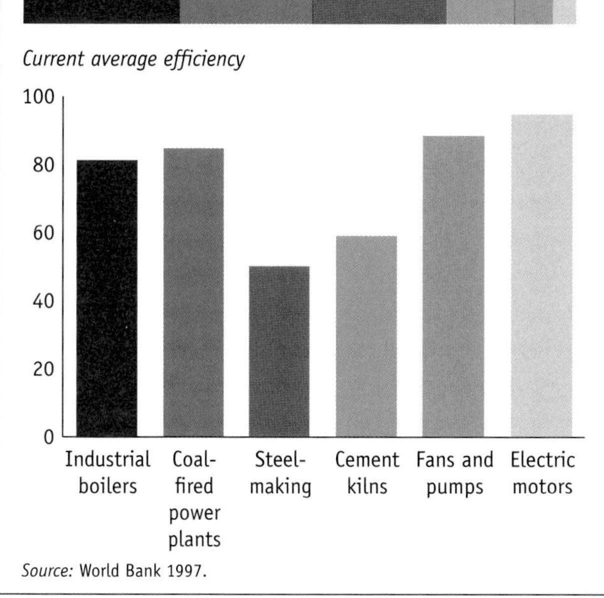

Source: World Bank 1997.

Planning and regulating effectively

Effective regulation and far-sighted planning can also promote a cleaner and healthier future for China. The main regulatory imperative is to phase out lead in gasoline. Beyond that, standards for vehicular emissions are still well below those in industrial countries, and should be raised. An automobile operating in China emits thirty to forty times as much carbon monoxide and forty to sixty times as many hydrocarbons as a car in the United States.

Second, far-sighted planning for urban land use could do much to shape a cleaner future for China's cities. High urban densities and low automobile ownership provide a unique window of opportunity for developing efficient and environmentally sustainable cities. To achieve this goal, urban planners will have to use a full range of instruments, from time-of-day tolls to parking restrictions. These efforts could be complemented by a policy that progressively relocates urban industries to suburban industrial parks, reducing pressures on city centers. The long-term effectiveness of such an approach has been demonstrated by Curitiba, Brazil, which began its now-famous urban planning twenty years ago. The city's gasoline consumption per capita is now 30 percent less than in many other Brazilian cities.

A third priority is to broaden the framework for regulation and monitoring. At present, the National Environmental Protection Agency monitors the emissions of only 70,000 firms, representing only a small fraction of the hundreds of thousands of industrial enterprises in China. The authorities find it difficult to effectively regulate millions of small township and village enterprises, and in certain sectors such as coking, metallurgy, and coal mining these enterprises have become major polluters. The recent decision of the State Council to close 60,000 highly polluting small township enterprises is a promising step.

A final priority is to reconsider the investment planning system, which currently does unintended environmental harm. At present, provinces face caps on the size of investment projects that can be undertaken without central government approval, so they tend to invest in inefficiently small power plants. About 80 percent of

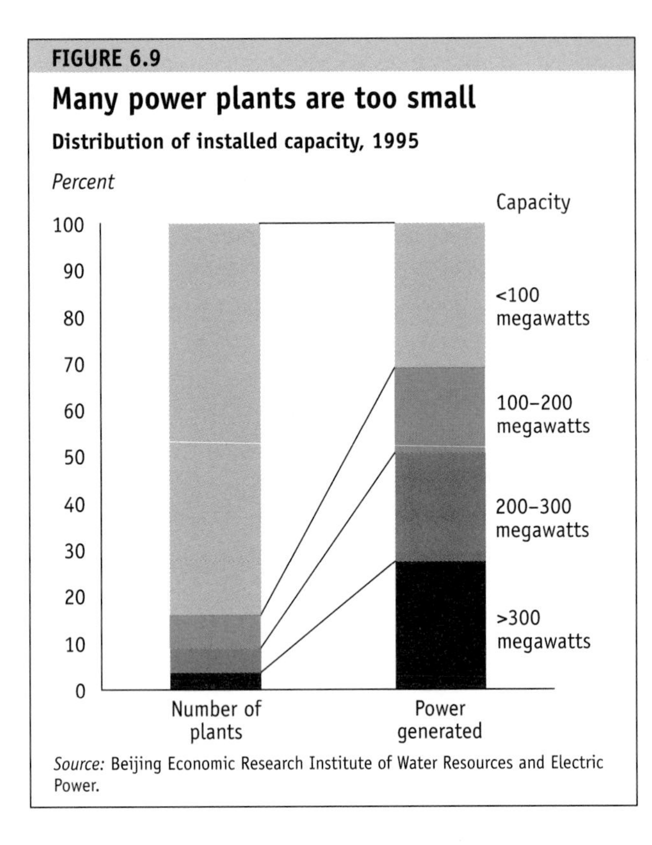

FIGURE 6.9

Many power plants are too small

Distribution of installed capacity, 1995

Source: Beijing Economic Research Institute of Water Resources and Electric Power.

China's power plants have a capacity below 100 megawatts, with lower energy efficiency and higher pollution abatement costs than larger plants (figure 6.9).

Community participation could help improve the regulatory process and reduce pollution. International experience shows that community pressures can be a powerful force for reducing pollution, even in the absence of formal environmental regulation.[16] Community participation does have two potential drawbacks, however. Community pressures tend to rise with incomes, which may encourage pollution-intensive industries to relocate to poorer areas with less opposition. And community pressures tend to focus on visible and prominent cases of pollution and may ignore less visible but equally harmful pollutants.[17]

In summary, a cleaner future is well within China's reach. Well-designed policies that harness market forces, encourage investment in alternatives, and use China's strong administrative and regulatory capacity will produce a cleaner environment for future generations. As with all reforms, these efforts will involve both public and private adjustment costs in the short term. But these costs pale in comparison with both the costs of inaction and the improvements in the quality of life that a cleaner environment brings (table 6.2).[18]

TABLE 6.2

Pollution abatement is a good investment

(returns to investment in pollution abatement technologies under the "deepening reforms" scenario)

	Required expenditures (percent of GDP)	Ratio of benefits to costs (percent): Conservative estimate of cost of pollution	Ratio of benefits to costs (percent): Moderate estimate of cost of pollution
Air pollution (excluding lead)	2.1	117	336
Water pollution	1.1	236	236
All forms of pollution	3.1	114	285

Note: Required expenditures refer to the annualized costs of investments in pollution abatement, including operating costs.
Source: World Bank staff estimates.

Notes

1. This chapter draws heavily on World Bank (1997).
2. National environmental policies and regulations are formulated by the State Environmental Protection Commission, which reports directly to the State Council. The commission is supported by the independent National Environmental Protection Agency, which disseminates environmental policy and regulations, collects environmental data, provides training and support to local environmental protection bureaus, and advises the commission on environmental policy. Pollution control policies are built on three fundamental principles: prevention first, then prevention with control; polluters pay; and a strong regulatory framework. These principles are given voice through a range of regulatory policies, including environmental impact assessments, pollution levies and discharge permits, and mandatory pollution control programs.
3. World Bank (1997, chapter 5).
4. China National Environmental Protection Agency (1996, pp. 7–23).
5. The World Health Organization (WHO) recently eliminated its standard for total suspended particulate emissions because there is no identifiable threshold below which health impacts are negligible.
6. Data on the relative contributions of various sources of lead in air, soil, and water are not available for China. In addition to automobiles, mining, and lead smelting, the production and recycling of batteries and cables are believed to be the primary sources.
7. These estimates are drawn from a study in Chongqing that drew on cross-sectional data on health and air quality in different zones of the city. An increase of 100 micrograms per cubic meter of particulates was associated with a 40 percent increase in mortality rates from chronic obstructive pulmonary disease, while an increase of 100 micrograms per cubic meter of sulfur dioxide lead to a 23 percent increase in mortality. See World Bank (1997).
8. World Bank (1997).
9. In 1990 four water-related diseases (diarrhea, hepatitis, trachoma, and intestinal worms) accounted for 1.5 percent of total deaths. These diseases cost 0.01 disability-adjusted life-year per capita in China, compared with 0.02 in Latin America, 0.04 in India, and 0.06 in Sub-Saharan Africa (World Bank 1993).
10. World Bank (1997).
11. Box 6.1 provides estimates of the costs of various sources of power, including wind power. Somewhat more optimistic estimates of the economic viability of wind power may be found in World Bank (1996). Also see Johnson and others (1996, pp. 28–30).
12. Grossman and Krueger (1995).

13. China State Council (1994, chapter 4). Environmental national accounts are intended to augment standard national accounts by deducting the costs of environmental degradation from measures of national income and the depletion of nonrenewable natural resources from measures of national wealth.

14. In 1995 steam coal delivered to noncoal-producing coastal cities (inclusive of the value added tax) cost about $40 a ton at market exchange rates. The spot price of Australian coal exports fluctuated between $38 and $48 a ton.

15. World Bank (1997).

16. Hartman, Huq, and Wheeler (1997).

17. Dasgupta and Wheeler (1997).

18. The returns to investment in pollution abatement presented in table 6.2 represent conservative estimates, as they do not take into account the effect of future relative price changes on the valuation of costs and benefits. If these are taken into account, the returns are substantially higher.

References

China National Environmental Protection Agency. 1996. *Selected Documents from the Fourth National Environmental Conference.* Beijing: Environmental Science Press.

China State Council. 1994. *China's Agenda 21: White Paper on China's Population, Environment and Development in the 21st Century.* Adopted at the 16th executive meeting of the State Council. Beijing.

Dasgupta, S., and D. Wheeler. 1997. "Citizen Complaints as Environmental Indicators: Evidence from China." Policy Research Working Paper 1704. World Bank, Policy Research Department, Washington, D.C.

Grossman, G., and A. Krueger. 1995. "Economic Growth and the Environment." *Quarterly Journal of Economics* 110(2): 379–406.

Hartman, R., M. Huq, and D. Wheeler. 1997. "Why Paper Mills Clean Up: Determinants of Pollution Abatement in Four Asian Countries." Policy Research Working Paper 1710. World Bank, Policy Research Department, Washington, D.C.

Johnson, T., J. Li, Z. Jiang, and R.P. Taylor, eds. 1996. *China: Issues and Options in Greenhouse Gas Emissions Control.* World Bank Discussion Paper 330. Washington, D.C.

World Bank. 1993. *World Development Report 1993: Investing in Health.* New York: Oxford University Press.

———. 1996. "China: Renewable Energy for Electric Power." Report 15592-CHA. China and Mongolia Department.

———. 1997. *Clear Water, Blue Skies: China's Environment in the New Century.* Washington, D.C.

chapter seven

Integrating with the World Economy

The opening of China's economy was an integral part of its economic reforms and a central element in its growth. Between 1978 and 1995 the value of exports and imports as a share of GDP tripled and China became the world's second-largest recipient of foreign direct investment (FDI) after the United States. The links between trade, FDI, and China's high savings rates have been a key factor in its rapid growth.[1]

These links will also be crucial for future growth. Deepening integration with the world trading system will bring further benefits of China's comparative advantages and provide clearer domestic signals on where to allocate resources. Increasing integration with the international financial system will help lower the cost of capital and deepen domestic financial markets. Sustaining FDI will bring new management know-how, the latest technologies embodied in capital equipment, and the most recent techniques in international marketing.

Accelerating the virtuous circle of trade and growth will require a firm timetable for reducing trade restrictions to acceptable international levels, supplemented by better procedures for resolving trade disputes. It will also require further improvements in the climate for FDI and observing international legal norms on the protection of property rights.

The government has committed itself to this path in the Ninth Five-Year Plan and Fifteen-Year Perspective Plan. At the same time, the authorities recognize that further trade liberalization will require difficult and costly adjustments in parts of domestic industry: steel, machinery, chemicals, automobiles, and consumer electronics—all industries that tend to be dominated by state enterprises. These adjustments could temporarily increase unemployment. They may also weaken the banking system.

Serious though such consequences would be, they can be planned for and dealt with, and should not deter the authorities from continued trade liberalization. The long-term growth benefits of a more liberal trading system far outweigh the short-term costs of adjustment (which, as other countries have learned, are usually lower than was initially thought). Chinese industries will improve their international competitiveness significantly if they gain access to high-quality imports, especially capital equipment. Closer links with international partners and competitors will also be vital to long-term success in export markets. And the pressure of world competition will encourage Chinese firms to adopt best practices and capture scale economies.

China's international integration is important to the rest of the world too. Its trading partners will benefit from exporting to its vast and largely untapped markets. More intensive trade will benefit all, just as a rising tide raises all ships. Moreover, foreign investors will be interested in China's sea of skilled and disciplined workers. Over the next several decades industrial countries will likely have unprecedented surpluses of long-term private capital seeking a safe and productive haven. Marrying this surplus capital to China's surplus labor could unleash productive power of enormous significance in the twenty-first century, to the mutual benefit of China and the world.

Deepening trade integration

During the late 1970s China's imports and exports accounted for about 13 percent of GDP—one of the lowest ratios in the world. Since then China's trade has surged to more than 30 percent of GDP (figure 7.1).[2] In U.S. dollar terms, trade increased nearly tenfold between 1978 and 1995, from $36 billion to $300 billion. As a result China is now the world's tenth-largest trading nation, accounting for 4 percent of world trade.

There are four noteworthy features of this change. First, China has merely caught up with global norms following a long period of autarky (figure 7.2). Large countries naturally trade less across their borders than smaller countries, precisely because they contain a wider diversity of resources offering opportunities for trade within their borders.

FIGURE 7.1

Outward and upward

(Trade–GDP ratios in 1987 prices)

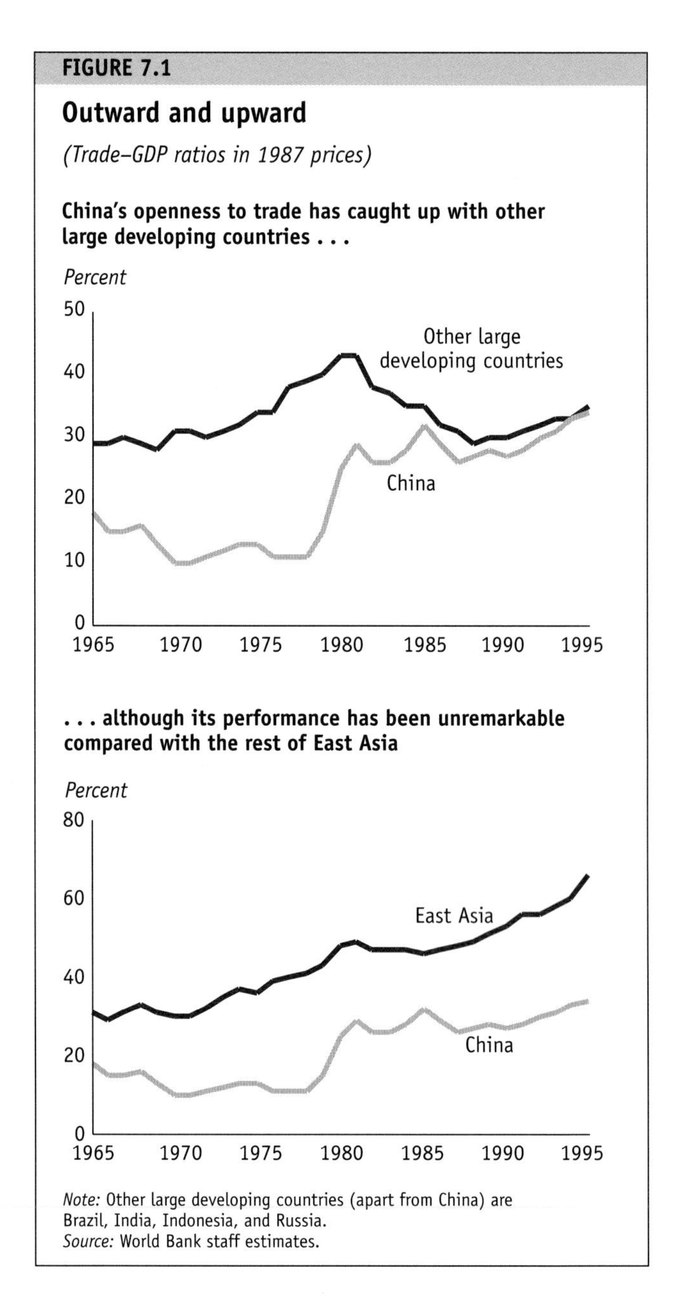

Note: Other large developing countries (apart from China) are Brazil, India, Indonesia, and Russia.
Source: World Bank staff estimates.

Second, these indicators of openness are somewhat deceptive. Roughly half of China's trade is imports processed into exports (usually by foreign-financed firms) and has little effect on the domestic economy; for example, it exerts few competitive pressures on state enterprises that sell their output in the domestic market. If the figures on trade are adjusted to exclude processing trade, then China is much less open than the global norm, even for a country of its size.

Third, China's external trade has been relatively diversified, in terms of both the commodities it trades and in its trading partners (figure 7.3). Exports are dominated by a wide variety of labor-intensive manufactures, with primary products accounting for only a modest share of the total. As a result China has not experienced the vagaries of volatile commodity prices.

Fourth, Hong Kong has played a crucial role as China's window to the world. Since reforms started in the late 1970s, Hong Kong has done much to channel goods and capital in and out of China. As a result the two economies were closely integrated well before their political unification (box 7.1).

China's integration with the rest of the world has been remarkably swift. But the process has left China with a complex set of trade and foreign investment policies administered by a range of occasionally overlapping ministries and agencies (box 7.2).

Protecting domestic industry from imports

The Chinese authorities regulate imports through tariffs, quotas, and licensing. Tariff rates are comparable to those imposed by several other large developing countries (figure 7.4). At the November 1995 meeting of the Asia-Pacific Economic Cooperation (APEC), China announced that it would lower tariffs on a number of goods, including coal and gas, textiles, apparel, leather, and other light and heavy manufactures. The net effect in 1996 was to reduce the simple unweighted average tariff from 36 percent to 23.4 percent and the weighted tariff from 28.1 percent to 19.8 percent. At the November 1996 APEC meeting the authorities announced further cuts. There are still many tariff concessions and exemptions, however: exemptions for export processing covered 44 percent of merchandise imports in 1995 and 45 percent in 1996.

Turning to nontariff barriers, thirteen commodities are subject to import quotas. These commodities were selected on the grounds that excessive imports could hinder the development of domestic industry or threaten the balance of payments. In addition, the Machinery and Electronics Import and Export Office

FIGURE 7.2

After rapid growth, China's trade ratio has reached large-country norms

Trade volume as a percentage of GDP, average 1978–94

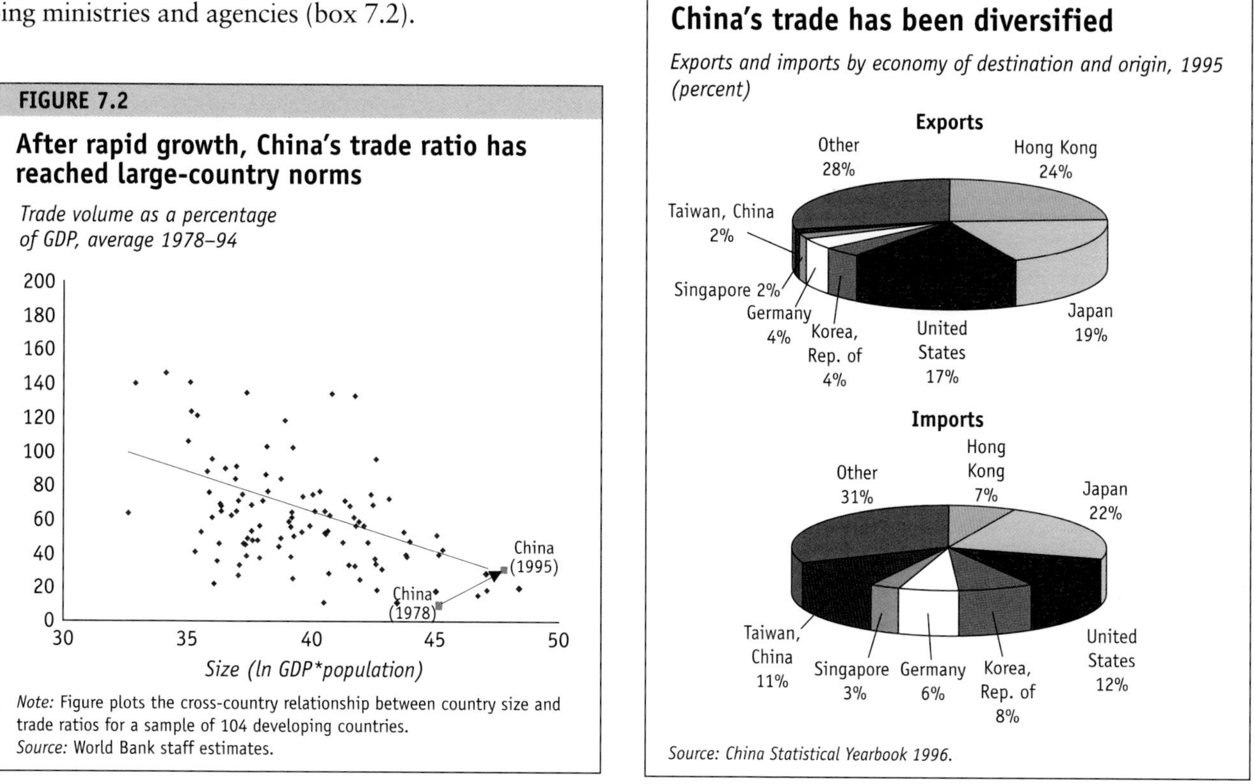

Note: Figure plots the cross-country relationship between country size and trade ratios for a sample of 104 developing countries.
Source: World Bank staff estimates.

FIGURE 7.3

China's trade has been diversified

Exports and imports by economy of destination and origin, 1995 (percent)

Source: China Statistical Yearbook 1996.

BOX 7.1

Hong Kong and mainland China: Ties that bind

Even after unification with China, Hong Kong's economic performance remains unshaken. Instead of suffering from capital flight and a brain drain, as some had feared, Hong Kong continues to flourish with a net influx of capital and talent. The Sino-British Joint Declaration and the Basic Law provide a unique "one-country, two-system" framework within which Hong Kong will remain a major trading and financial center. Over the past few decades Hong Kong and China's complementary strengths have knitted them together tightly. There is no reason these forces should not continue to deepen integration in the coming decades.

Hong Kong under the Basic Law

The Basic Law passed by the National People's Congress in 1990 stipulates that Hong Kong will retain its free market system for fifty years after unification. According to this law, the Hong Kong Special Administrative Region will:

- Protect private property rights, private ownership of enterprises, and foreign investments.
- Manage independent public finance, enact its own tax laws, maintain exclusive use of its financial revenues, and strive to achieve fiscal balance.
- Maintain an autonomous monetary system and safeguard the free operation of financial business and the free flow of capital.
- Retain the Hong Kong dollar as legal tender, ensure full convertibility, and manage the exchange fund independently.
- Pursue free trade policy, maintain the status of a free port, operate as a separate customs territory, participate in international organizations and trade agreements as a separate entity, and enjoy exclusively the export quotas and tariff preferences it obtains or makes.

The two monetary authorities further announced that:

- The two monetary systems and authorities will be mutually independent. The Hong Kong dollar will remain linked to the U.S. dollar at a fixed exchange rate.
- Both economies will treat the other's financial institutions as foreign. Mainland financial institutions operating in Hong Kong must abide by Hong Kong's laws. All financial claims and liabilities between the mainland and Hong Kong will be dealt with in accordance with internationally accepted rules and practices.
- The People's Bank of China will support Hong Kong's currency stability but will not draw on or resort to Hong Kong's exchange fund under any circumstances.
- Shanghai and Hong Kong will develop as complementary financial centers.

Integrated economies

The economies of China and Hong Kong are already intertwined through trade, investment, and personal contacts. This integration not only has spurred growth on the mainland but also has contributed to Hong Kong's transformation into a sophisticated center for international trade and financial services.

- China accounted for 37 percent of Hong Kong's foreign trade in 1996, while Hong Kong accounted for 47 percent of China's trade.
- Hong Kong is an entrepot trade and transshipment center for China, providing vital transportation, storage, insurance, packaging, and processing services. The profits from these activities, estimated at 25 percent of the value of the merchandise have boosted the growth of Hong Kong's service sector.
- Hong Kong is the main conduit of foreign capital for China. Since 1978, 57 percent of FDI in China has come from or been funneled through Hong Kong. About 90 percent of syndicated loans for the mainland are arranged in Hong Kong. By the end of 1996 Hong Kong's banking system owed $39 billion to banks in China, and had claims on Chinese banks and nonbanking entities of $46 billion.
- China is the second largest "foreign" investor in Hong Kong. By the end of 1994 China's investments in Hong Kong totaled more than $12 billion. More than fifty mainland companies ("red chips") have become pillars of Hong Kong's local establishment, and their market capitalization on Hong Kong's stock exchange has reached $24 billion. The listing of Chinese firms on the domestic stock market (B shares) and on the Hong Kong Stock Exchange (H shares) increased the financial integration of the two economies.
- Some 20,000 trucks travel between Hong Kong and the mainland every day, and Hong Kong residents made 29 million trips to China in 1996. More than 2 million mainlanders visited Hong Kong, and the immigration quota has been increased to more than 50,000 a year.

Complementarities and prospects

China and Hong Kong have a lot to offer each other. The service sector dominates Hong Kong's economy, accounting for 84 percent of GDP and employing 71 percent of the labor force (see figure). Hong Kong has highly efficient transport and telecommunications systems. Its deep water harbor and modern transport system mean it can support China's trade and financial development as a supplement to the mainland's overstretched infrastructure. Hong Kong, however, suffers from labor and land constraints, which make its populous neighbors—Shenzhen and Guangdong—so attractive to Hong Kong capital.

Especially valuable is Hong Kong's considerable expertise in finance, trade, and public administration—skills that China sorely needs. Hong Kong is home to 405 banks and branch offices, 335 of them foreign. The net daily turnover of its foreign exchange market ranks fifth in the world, and capitalization of its equity

BOX 7.1, continued

Hong Kong and China: Ties that bind

Hong Kong's service sector has a lot to offer

Percentage of GDP (1995)

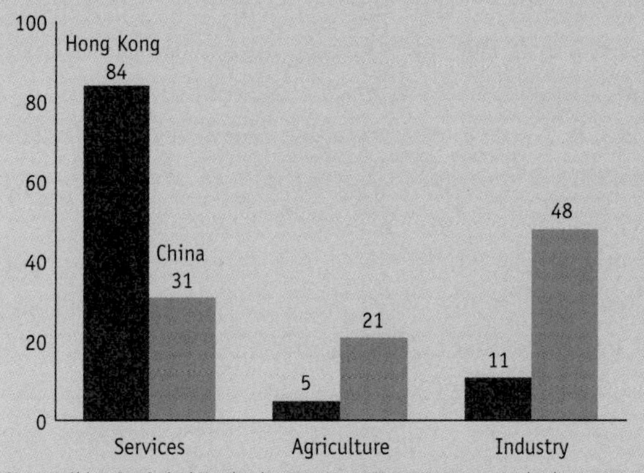

Source: China Statistical Yearbook 1996; Hong Kong Monetary Authority 1997a.

Tapping Hong Kong's human capital, 1995

Category	Hong Kong	China
Professional staff (per 100,000)		
Accountants	35[a]	4
Lawyers	62[a]	7
Share of labor force (percent)		
Trade and commerce	44	12
Finance and insurance	16	2
Transport and communications	7	6
Services (total)	71	24

a. Data are for 1996 or March 1997.
Source: Hong Kong Monetary Authority 1997b. *China Statistical Yearbook 1996.*

market ranks ninth. Hong Kong has about 1,300 accounting firms and 3,000 management consulting firms (see table). It is the regional headquarters of nearly 800 foreign companies. This concentration of professional skills could benefit China's financial sector reform and development.

Hong Kong has established a responsive, lean government and sound legal system. Information is freely accessible and the regulatory system transparent—factors that have been crucial to its success. It is in the interests of both China and Hong Kong to maintain Hong Kong's distinctive features. At the same time, being part of China will provide Hong Kong with a competitive edge over its strong Asian rivals.

Hong Kong's return to Chinese sovereignty will have equally profound consequences for the mainland economy. Hong Kong, as China's window to the world, will become an increasingly important source of knowledge and expertise for China in private and public sector management.

BOX 7.2

Who's who in China's trade policy

Under the State Council, eight major institutions with overlapping responsibilities formulate and administer trade policy.

Policy formulation. The State Planning Commission and the Ministry of Foreign Trade and Economic Cooperation jointly formulate national foreign trade policies. The ministry negotiates international trade agreements.

Quota and license allocation. The State Planning Commission allocates import and export quotas for goods that are deemed essential to China's economy and people's livelihood. The Machinery and Electronics Import and Export Office coordinates and administers imports and exports of principal machinery and electronic products. The Ministry of Foreign Trade and Economic Cooperation issues import and export licenses.

Tariff and nontariff barriers. The Customs Tariff Commission, reporting to the State Council, sets tariffs. The Customs General Administration administers the trade regime, including tariffs and nontariff barriers. The Ministry of Foreign Trade and Economic Cooperation also has direct responsibility in this area.

Direct trading rights. The Ministry of Foreign Trade and Economic Cooperation grants trading rights to enterprises that meet certain criteria. Foreign trade corporations, under the ministry's administration, are allowed to trade in all products except for a small number that can be handled only by designated foreign trade corporations.

Foreign exchange control. The State Administration of Foreign Exchange, which reports to the People's Bank of China, is responsible for carrying out foreign exchange policies.

applies quotas to fifteen machinery and electronic products, such as automobiles and refrigerators.[3]

Automatic import registration covers a wide range of imports, including oil, nonferrous metals, polyester, and cotton. Some of these imports are subject to quotas and licenses, and some can be traded only through state trading monopolies. Although automatic import registration is not intended as a nontariff barrier, it risks having the same effect. Importers must demonstrate a market need for the goods and prove that they can pay for them. But since firms would not attempt to import unless they had decided that imports would meet their needs better than domestic supplies, it is not clear why a government department is better qualified to make this judgment than the managers of the importing firm. The financing requirement appears similarly redundant—if the firm does not have the finance, it will not import.

Of course, effective rates of protection differ significantly from nominal tariffs. The difference is particularly pronounced for machinery and equipment (figure 7.5).[4]

Reducing protection

China's eventual accession to the World Trade Organization (WTO) will result in further reductions in tariffs. If its current formal WTO offer is implemented, weighted average tariffs will fall from 19.8 percent today to 16.2 percent by 2005.[5] These reductions will particularly benefit manufactures exporters such as Japan, Korea, and the European Union, which as recently as 1995 faced tariffs ranging from 30 to 40 percent. Raw material exporters to China will gain less, primarily because the tariffs they face are already lower.

WTO accession will also involve big reductions in nontariff barriers for China, eliminating almost all except those on food and certain other primary products (figure 7.6).

If China implements its WTO offer and industrial countries, in turn, abolish quotas under the Multi-Fibre Arrangement (MFA), the welfare gains to China by 2005 could reach $116 billion a year, and the benefits to the rest of the world could be twice as great (table 7.1). Lowering trade barriers will create new opportunities for increased specialization and gains from trade. Reducing the variability of tariffs across sectors will ease distortions—their standard deviation will fall from 23 percent in 1995 to 12 percent in 2005. Furthermore, greater transparency in trade rules will reduce distortions and eliminate unproductive rent seeking.

At the same time, however, the authorities need to anticipate the domestic consequences of further import liberalization, which are likely to be concentrated in capital-intensive, scale-sensitive industries where profits are high (figure 7.7).[6] Such industries include steel, automobiles, machinery, electronics, and heavy chemicals. Many of these industries are concentrated in regions where unemployment is already high. Moreover, any financial difficulties they may have

FIGURE 7.4
Declining tariff barriers

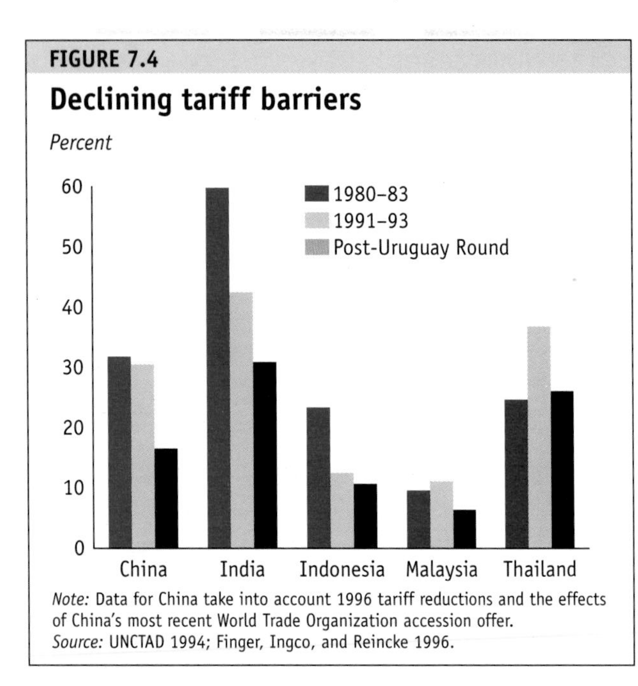

Note: Data for China take into account 1996 tariff reductions and the effects of China's most recent World Trade Organization accession offer.
Source: UNCTAD 1994; Finger, Ingco, and Reincke 1996.

FIGURE 7.5
Nominal and effective rates

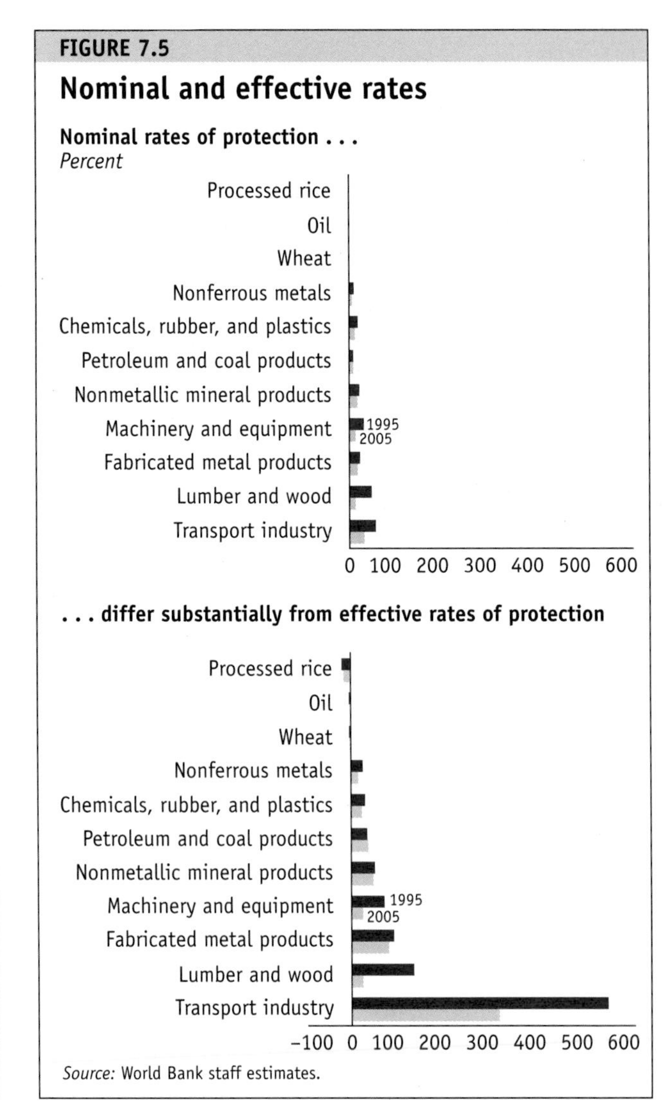

Source: World Bank staff estimates.

would quickly affect the fragile financial system (see chapter 3). A lack of data prevents an accurate assessment of the possible shock, but it could be much larger than the current portfolio problems of state commercial banks.

Should the prospect of these difficulties deter the authorities from liberalizing trade? No—for two reasons. First, experience in other countries has shown that the costs of adjustment, while significant, are usually less than was first thought. Economies tend to adjust positively and rapidly to steady, determined reductions in trade barriers. In the long run the gains from trade liberalization far outweigh the adjustment costs. Around the world, there is growing consensus among policymakers and academics that trade liberalization is one of the key ingredients for higher growth. China's experience over the past two decades attests to this view.

FIGURE 7.6

Proposed nontariff barriers under World Trade Organization accession

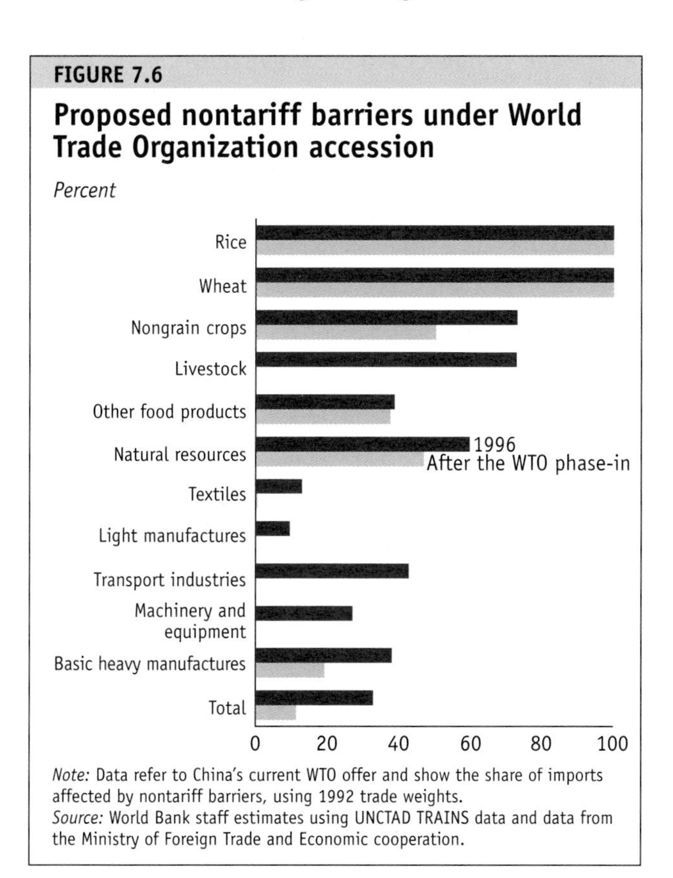

Note: Data refer to China's current WTO offer and show the share of imports affected by nontariff barriers, using 1992 trade weights.
Source: World Bank staff estimates using UNCTAD TRAINS data and data from the Ministry of Foreign Trade and Economic cooperation.

TABLE 7.1

Projected welfare gains from trade liberalization
(average annual benefit, in billions of 1992 dollars)

Economies	With WTO offer	With WTO offer and elimination of MFA quotas
China	83	116
Rest of the world	340	332
European Union	71	81
United States, Canada, and Mexico	38	44
Japan	61	62

Source: World Bank staff estimates.

FIGURE 7.7

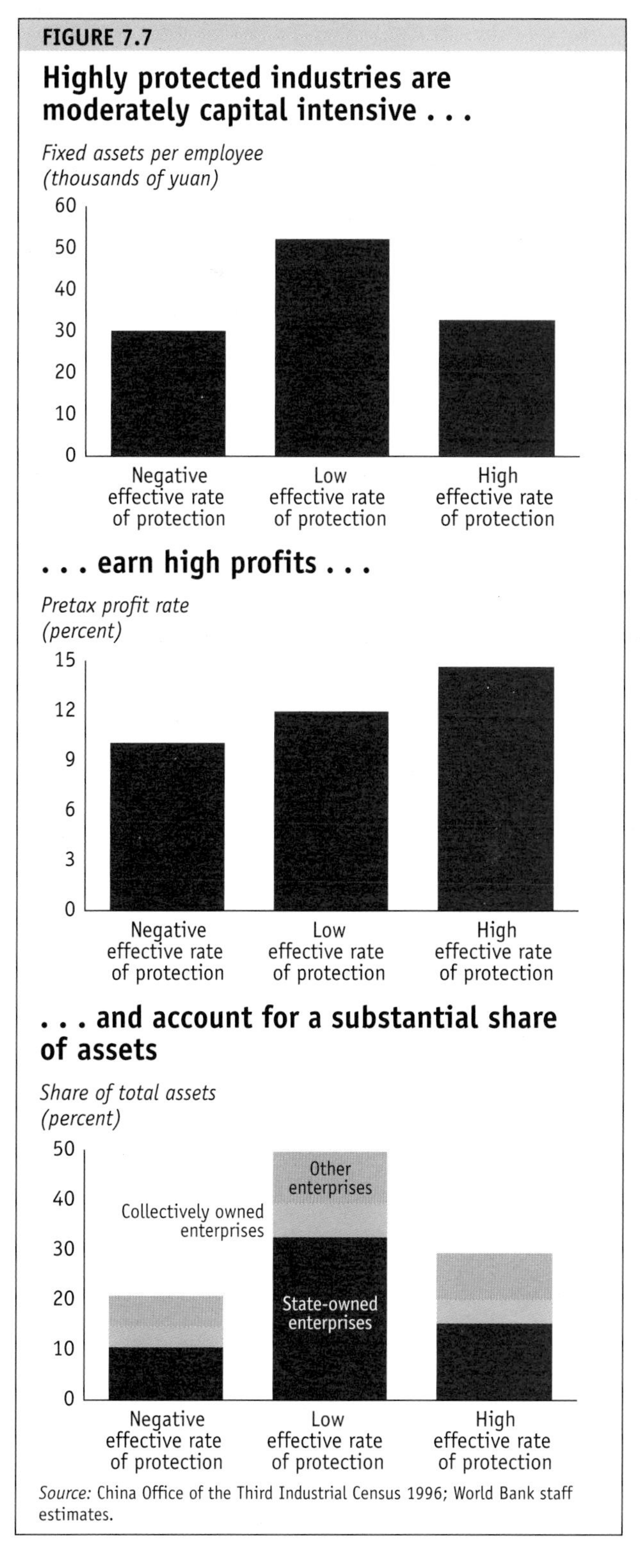

Source: China Office of the Third Industrial Census 1996; World Bank staff estimates.

Second, the government can prepare plans to deal with the adjustment costs. One way is to continue reforming banks, state enterprises, and labor markets (see chapters 3 and 4). The more that is done in these areas, the easier it will be for the economy to adjust to the changes in relative prices brought on by trade liberalization. In addition, the government will have to finance some of the adjustment costs through unemployment benefits, payments to pensioners of failed enterprises, and recapitalization of state banks (but only once they are fully commercial). Meeting these costs will require further efforts to collect taxes, reduce the budget deficit, and shift government spending toward social investments and measures that support reforms.

In any case, considerable efficiency gains can still be wrung from further trade liberalization. Developing countries such as China can boost productivity by importing a larger variety of intermediate products and capital equipment embodying foreign technology.[7] Since 96 percent of the world's research and development is done in industrial countries, their technical progress can be captured relatively cheaply by liberalizing imports of such goods. More important, imported technology tends to spread rapidly across the domestic economy (see chapter 3). The productivity benefits for domestic investment are sometimes several times greater than the benefits acquired directly through the imports themselves.

Deepening financial integration

China's rapid growth in trade has been accompanied by big increases in capital inflows (figure 7.8). A World Bank index shows that China's financial integration with the world economy has grown sharply since the mid-1980s.[8] China now accounts for 40 percent of FDI to developing countries and is the largest recipient of FDI after the United States. It also receives 15 percent of new cross-border commercial debt acquired by developing countries, and has begun to tap the growing pool of portfolio investment.

Foreign direct investment

Before 1979 China attracted almost no FDI. By 1995 net inflows had reached $38 billion, accounting for 13 percent of domestic investment, 13 percent of industrial output, 12 percent of tax revenues, and 16 million jobs. Much FDI has been in special economic zones, underpinning the rapid growth of trade. In 1995 one-third of China's exports and half of its imports involved joint ventures between Chinese and foreign partners.

The bulk of FDI inflows have come from the Chinese diaspora in Hong Kong, Singapore, and Taiwan, China. In addition, some of the inflows have been domestic capital that has "roundtripped" its way through Hong Kong and back to the mainland to take advantage of the tax privileges available to foreign investors.[9] FDI goes largely to coastal regions—the nine coastal provinces and three municipalities have consistently attracted more than 85 percent of the total—and has been concentrated mainly on tradable manufactures. Recently, however, the range of investing countries has widened, and some FDI has been going into import-substituting activities.

Over the long term sustaining FDI and improving its efficiency will require China to face three challenges. First, China will need to maintain macroeconomic stability, which international evidence indicates is the most important determinant of FDI inflows. Stability should be supplemented by efforts to reform the regulatory and institutional framework for FDI, especially at the provincial and county levels, and to enhance the transparency of tax and foreign exchange rules.

Second, China must weigh the benefits and costs of incentives to promote FDI. China recently unified the tax regime for foreign and domestic enterprises, elimi-

FIGURE 7.8

A tidal wave of capital inflows

Billions of U.S. dollars

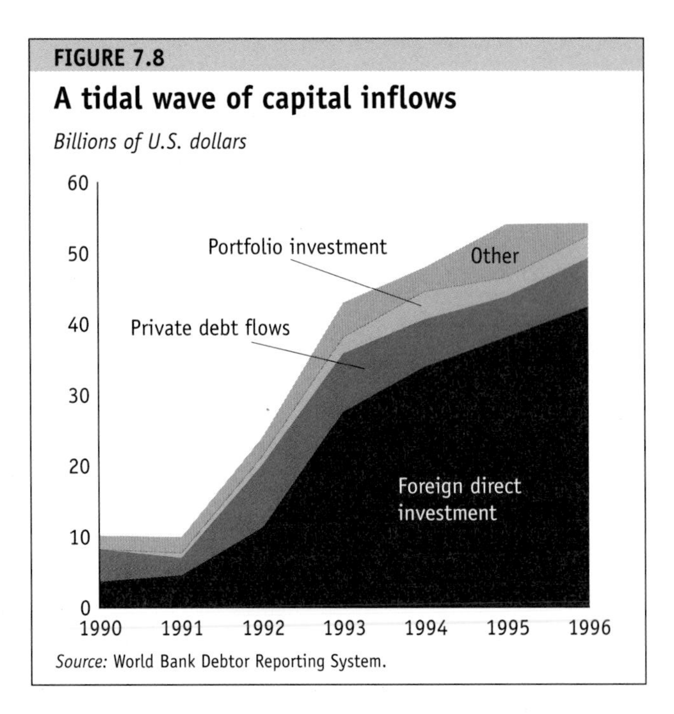

Source: World Bank Debtor Reporting System.

nated duty exemptions on imported capital goods, and restricted tax incentives from provinces and cities. These steps could slow the growth of FDI because they could discourage roundtripping. They are unlikely to deter genuine FDI, however, because the overwhelming evidence worldwide is that tax and other incentives have only marginal effects on the location decisions of firms.

Third, the government wants to shift FDI toward infrastructure (especially in the interior) and away from manufacturing and real estate. Foreign investors typically have shied away from infrastructure projects, partly because they have found themselves negotiating with both the central government and the provinces and the procedures are inconsistent from one project to the next. In addition, investors have not received assurance on performance obligations from their Chinese counterparts, and have been deterred by China's underdeveloped legal and regulatory system.

Many of these issues are being resolved. The legal framework is being developed. Pricing reform, especially for utilities, has made power investments more attractive. New financing techniques are available, often offered in conjunction with multilateral financial institutions such as the World Bank.

As a result the number of FDI transactions in infrastructure is increasing (see chapter 3). Foreign investors are involved in a dozen power generation projects, each with a capacity of over 100 megawatts, as well as in ports, highways, and railways. Several provinces are packaging publicly financed assets into ventures, selling shares to foreign investors, and using the proceeds to finance new projects. In China, this is known as "cascading finance." It might be used to finance a series of dams being planned on the Yellow River (among other projects), though this would require establishing a comprehensive river basin authority to coordinate the involvement of the provinces.

External debt

Foreign commercial lending has expanded as market perceptions of China's creditworthiness have improved. By the end of 1996 external debt totaled about $130 billion, with both the debt to exports ratio (85 percent) and the debt to GNP ratio (20 percent) standing at less than half the developing country average and among the lowest in Asia. Even the currency composition of China's debt has been diversified.

To ensure that its external liabilities remain managable and that it maintains easy access to international capital markets, China needs to continue displaying prudence in its foreign borrowing program. In the past, foreign borrowing bolstered foreign exchange reserves. But these reserves now cover more than eight months of imports. In the near future, China does not need to borrow much.

Even within this conservative framework, however, fundamental reforms are still possible. For example, before state enterprises are allowed to borrow abroad on their own account, they should be subject to certain new criteria, including a modern corporate structure in accordance with the Company Law, use of the new accounting system, accounts audited according to internationally accepted standards, profitable operations over the previous three years, demonstrated ability to service foreign debt, and so on.

Over the long term three other issues merit careful consideration by the government:

- Monitoring short-term debt more effectively, through timely and thorough supervision of the commercial banks by the People's Bank.
- Developing a clear framework for foreign borrowing for infrastructure, including the use of selective performance guarantees. The World Bank could assist this effort through its expanded guarantee program.
- Integrating external debt management with a public finance framework that considers the *total* financing requirement (domestic and external) of the consolidated public sector. As other countries have shown, this may require the government to rearrange the way that external debt management is coordinated between ministries.

Portfolio flows

The potential for portfolio investment is huge. So far, inflows into China have been equivalent to less than 0.5 percent of GDP; the sixteen developing countries that attracted most developing country portfolio flows in 1995 averaged about 2 percent of GDP.

Portfolio flows into China are limited to buying those equities that can only be subscribed by and traded among foreigners (including Hong Kong residents).[10] These are the designated B shares on the Shanghai and Shenzhen stock exchanges; A shares are reserved for Chinese residents. B shares, introduced in 1992, have attracted about $2 billion of portfolio investment in sixty stocks so far. In addition, some Chinese companies are listed on the Hong Kong and New York stock exchanges.[11] By 1995 about $4 billion in equities had been issued in overseas markets by Chinese companies.

The segmentation of China's stock market between domestic and foreign investors was introduced to assuage concerns about the volatility of private capital flows. Mexico's 1995 peso crisis reinforced this cautious approach. Because the B share market is relatively illiquid (table 7.2), the shares tend to sell at a discount, and arbitrage opportunities generate some illegal transactions. Nevertheless, B share prices are poorly correlated with A share prices, suggesting that the segmentation is broadly effective.

Intensifying financial integration

Financial integration with the rest of the world is still unfolding in China. At the same time, technological progress, financial innovations, and deregulation have spurred considerable private inflows. Not all of these inflows are legal. For example, Hong Kong dollars are widely used in parts of southern China, and sizable capital flows are rumored to move between the mainland and Hong Kong each day. China's official figures show a large errors and omissions component in the balance of payments, a sign of heavy unrecorded movements.

So integrating China financially with the world economy is not a choice for policymakers to make. Markets are making it for them. The issue for the government is how to manage the forces of integration for the overall benefit of the economy, and how to prevent the virtuous cycle of good policies and high growth from being damaged by destabilizing capital flows.

Moreover, international private capital flows have grown explosively in recent years, responding to cross-border opportunities and driven by deregulation in many countries. Although FDI is still the biggest component of private capital flows, cross-border sales of bonds and equities have increased more sharply in recent years, and are now 30 percent of the total.[12]

This worldwide expansion presents great opportunities for China. For example, competition from foreign banks could promote the efficiency of domestic banks. Similarly, portfolio inflows could increase liquidity in domestic capital markets and encourage improvements in market infrastructure and the regulatory framework.

But China must overcome numerous hurdles to achieve the preconditions for full financial integration and currency convertibility on the capital account. International experience shows that unrestricted capital inflows can strain macroeconomic stability. Moreover, a weak financial sector magnifies the risks associated with capital inflows if competition from foreign banks prompts domestic banks to lend to less creditworthy clients. Domestic banks are also exposed to new instruments, such as derivatives, with which they have little experience.

On balance, the benefits and risks of financial integration suggest a cautious approach to liberalizing capital flows. But procrastination is not the answer, since the forces of markets and technology will bring unrelenting pressure for cross-border capital flows. It is better to actively harness these forces to promote the economy and the overall soundness of the financial system than to allow government inactivity to spawn unpredictable and potentially destabilizing movements of capital.

TABLE 7.2
The B shares market is less liquid
(turnover as a share of market capitalization)

Share type	1995	1996
A shares	9.18	27.71
B shares	3.93	9.56

Source: China Security Regulatory Commission data.

Sharing the benefits of China's growth

By 2020 China's exports are projected to be nearly 10 percent of the world's total, ahead of Japan but still behind North America and the European Union.[13] This rapid integration with the world economy will have different—but mostly favorable—effects on the rest of the world.

Industrial countries will benefit unequivocally from China's rising demand for capital- and knowledge-intensive manufactures and primary products and from significant terms of trade gains. Only their labor-intensive manufacturing will face a squeeze, but much of that structural change would have happened anyway, with or without China.

Among developing countries the effects of China's integration will depend on how much they trade directly with China and how closely they compete in third markets. Countries that trade heavily with China but are not close competitors (such as Korea) will likely gain substantially. Low- and middle-income countries that are close competitors (India, Indonesia, the Philippines, and Thailand) will experience some terms of trade losses on their exports of labor-intensive manufactures. But their total trade will keep growing, and perhaps their world market shares as well. Regions with fewer trade ties to China (Latin America, Sub-Saharan Africa, Eastern Europe, Central Asia) will neither make significant gains nor suffer major losses.

Effects on industrial economies

China's growth and internationalization should have three main effects on industrial economies. First, they will see faster growth in demand for capital- and knowledge-intensive products. The World Bank projects that their exports to China will grow at 8 percent a year, boosting overall export growth from about 2 percent a year (without China) to 2.5 percent. While most of this increase will involve knowledge- and capital-intensive products, exports of primary products (such as foodgrains) will also rise.

Second, this rapid growth in demand for industrial country exports will raise their prices relative to those of labor-intensive products. By 2020 the cumulative terms of trade gain will total 6.5 percent for Japan, 4.0 percent for North America, and 1.4 percent for Western Europe. Combined with the extra volume growth, the overall effect will be to enhance income growth in industrial countries.

Finally, many observers argue that global integration damages the jobs and incomes of unskilled workers in industrial countries. From this some conclude that China's vast size and the rapid growth of its trade will exacerbate the damage. Historical evidence suggests otherwise, however. In industrial countries most of the decline in employment in labor-intensive industries such as clothing occurred well before Chinese exports had significantly penetrated these markets (figure 7.9)

Effects on developing economies

China is expected to maintain its specialization in labor-intensive manufacturing and gradually lose market share in resource-intensive products. Compared with other major developing countries, China is poorly endowed with land and physical capital, and its stock of human capital (particularly in terms of years of higher education) is also lagging (table 7.3).[14] These weaknesses are likely to persist.

Looking at indexes of revealed comparative advantage across 129 industries for a set of large developing countries, China's pattern most closely resembles that of India, Indonesia, and Thailand.[15] Moreover, their trading structure has grown more like China's over the past decade.

In the future China's comparative advantage is expected to shift toward intermediate technology manufactures and away from low-technology clothing. By 2020 China is projected to gain 10 percent market share in light manufactures (leather, fabricated metal products, and miscellaneous manufactures) and 8 percent in

FIGURE 7.9

Employment in garment manufacturing in industrial economies declined before Chinese exports expanded

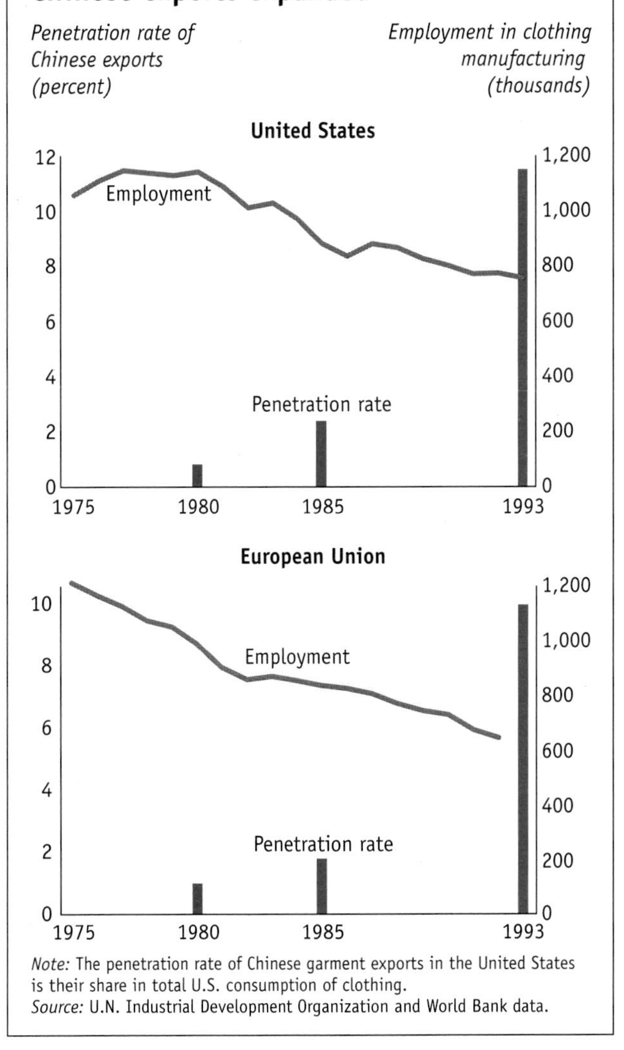

Note: The penetration rate of Chinese garment exports in the United States is their share in total U.S. consumption of clothing.
Source: U.N. Industrial Development Organization and World Bank data.

TABLE 7.3

Compared with most large developing economies, China is poorly endowed with land, capital, and educated workers

Country	Area under crops (hectare per worker)		Physical capital (thousands of U.S. dollars per worker)		Secondary education (years per person)		Tertiary education (years per person)	
	1995	2020	1995	2020	1995	2020	1995	2020
China	0.12	0.10	0.16	13.15	4.22	5.94	0.18	0.36
Brazil	0.61	0.58	9.56	31.22	1.76	3.83	0.12	2.08
India	0.30	0.29	1.68	8.20	3.92	4.38	0.45	1.30
Indonesia	0.19	0.18	2.70	21.94	2.34	6.20	0.51	1.59
Korea, Rep. of	0.07	0.06	21.48	115.17	5.65	6.44	3.03	6.75
Malaysia	0.42	0.40	15.84	139.62	3.78	5.70	0.53	2.20
Philippines	0.20	0.19	2.99	27.17	3.48	6.17	2.74	5.02

Source: Ahuja and Filmer 1995; World Bank staff estimates.

transport equipment and other machinery. Its fast-growing Asian neighbors—Indonesia, Malaysia, and Thailand, for example—are also expected to gain market share. In highly capital-intensive heavy manufactures (chemicals, rubber, plastics, paper, iron and steel, nonferrous metal) China is projected to increase its market share by 4 percent, but industrial countries will maintain their dominance.

As for clothing, China and Hong Kong could lose as much as 16 percent of the world market by 2020. Industrial countries are expected to lose an additional 19 percent, while India, Indonesia, the Philippines, and Thailand will have the biggest boost to market share. Neither Latin America nor Sub-Saharan Africa is likely to gain much from this shift in trading patterns. China is also projected to lose market share in some agricultural products—rice, other grains, livestock, and processed food. The gainers will be some industrial countries (Australia, Canada, the United States) and some Asian neighbors (notably India, Indonesia, Malaysia, and the Philippines).

The expected rapid rise in China's manufactured exports will tend to depress their relative prices. Meanwhile, its rising imports of machinery and transport equipment will make these products relatively more expensive. Thus the terms of trade changes for any country will depend on its specialization in these two categories. Economies that are ahead of China in industrial development (Korea, Malaysia, and Taiwan, China) could make moderate gains and suffer small losses in terms of trade. The same is broadly true of East Asian and South Asian developing countries that are in direct competition with China. Other countries—in Sub-Saharan Africa and Latin America—will make small gains and suffer only minor changes in their terms of trade (table 7.4).

To summarize, China's rapid growth and continued liberalization will be an opportunity for the world economy, not a threat. However, much will depend on whether China's largest export markets remain open. Rapid export growth worked for Japan and Korea in the 1950s and 1960s, but it may prove more difficult for China in the twenty-first century. Trade issues have become more sensitive and complicated. If China's trading partners were to raise barriers to its exports and China were to retaliate, the losses to China—and the rest of the world—could be large.

To avoid this, China's accession to the WTO becomes imperative. Not only will trade relations be easier to

TABLE 7.4

Projected changes in trade shares and terms of trade in developing economies

Economies	Change in world market shares for all sectors, 1992–2020 (percentage points)	Cumulative change in terms of trade, 1992–2020 (percentage points)
Asian economies that trade with China but are not major competitors[a]	1.0 to 3.0	-3 to -9
Asian developing countries with endowments similar to China[b]	0.3 to 2.8	-8 to -13
Other developing economies	-0.2 to 0.1	-3 to 4

a. Republic of Korea, Malaysia, and Taiwan, China.
b. India, Indonesia, Philippines, and Thailand.
Source: World Bank staff estimates.

handle within that context, China itself would also benefit from adopting international rules on competition and intellectual property rights, ensuring an open and transparent trade regime, reducing its import barriers, and resolving disputes through the neutral mechanisms of the WTO.

Notes

1. This chapter draws on World Bank (1997a).

2. Two measurement issues are of key importance in comparing the level and pace of China's integration with the world. First, as discussed in chapter 1, China's GDP may be understated by a considerable margin. Since these measurement issues have little effect on the measurement of trade, this implies that China's trade to GDP ratio may be considerably overstated. The second point relates to the growth of China's trade to GDP ratio. The figures cited in the text refer to the real growth of the trade to GDP ratio, measuring both variables in constant 1987 prices. But the increase in China's nominal trade ratio is even more rapid, from 10 percent in 1978 to over 45 percent in 1995. This more rapid rate of increase reflects primarily the sustained real exchange rate depreciation (that is, the increase in the relative price of traded goods) in China during this period.

3. The administrative procedures used by the Machinery and Electronics Import and Export Office to allocate these quotas appear to be extremely complex. Importers submit applications to their local administering bodies, which submit them to the office for approval. The office issues certificates of quotas that importers can use to obtain import licenses from the Minstry of Foreign Trade and Economic Cooperation. Customs clears the imports on presentation of the import licenses.

4. Calculation of the effective rates of protection used tariff and nontariff data from Beijing CDS Information Consulting Company, the China Ministry of Foreign Trade and Economic Cooperation, and the United Nations Conference on Trade and Development. The tariff equivalents of nontariff barriers were obtained by price comparisons using data from Unirule Institute (1996) and the International Comparison Programme. For goods protected by both tariff and nontariff barriers, the higher value was chosen. Input-output coefficients obtained from the 1992 Global Trade Analysis Project input-output table (Hertel 1997) were used to calculate the effective rates of protection.

5. The analysis was carried out using the Global Trade Analysis Project model and database.

6. For figure 7.7 effective protection rates for manufacturing sectors in the Global Trade Analysis Project model were mapped into manufacturing sectors reported in China Office of the Third Industrial Census (1996). Sectors with negative effective rates of protection are food processing, food production, beverages, tobacco, textiles, apparel, and leather and fur processing. Sectors with moderate effective rates of protection (less than 50 percent) are paper, petroleum, basic chemicals, chemical fibers, rubber, plastics, nonmetal products, ferro-alloy processing, and nonferrous processing. Remaining sectors have high effective rates of protection (more than 50 percent) and include wood processing, furniture, pharmaceuticals, metal products, machinery and equipment, transport equipment, and electronics. Assets refer to the net value of fixed assets. Profit rates refer to pretax profits as a share of net assets. Tobacco is excluded from all figures because of the unusually high profits reported in this sector (more than 100 percent).

7. See Coe, Helpman, and Hoffmaister (1995); Grossman and Helpman (1993); and De Long and Summers (1993).

8. The index uses a weighted average of three variables: country risk ratings, which measure a country's access to international financial markets; ratio of weighted private capital flows to GDP, which reflects a country's ability to attract private financing; and composition of private capital flows, which captures the diversification of inflows.

9. According to some estimates, "roundtripping" accounted for about 20 percent of recent FDI inflows (UNCTAD 1995).

10. Overseas investors are excluded from the domestic bond market.

11. New York Stock Exchange listings of Chinese company shares, called N shares, are in the form of American depository receipts (ADRs).

12. World Bank (1997b).

13. This section reports results from a large-scale computable general equilibrium global trade model. The assumptions behind the model are laid out in World Bank (1997a).

14. The Republic of Korea is the only country with less land per worker, but it is already more specialized in knowledge-intensive industries, with higher wages.

15. The revealed comparative advantage index is defined as the ratio of the share of a product in a country's total exports to the share of that product in world exports. A value greater than one indicates that a country has a revealed comparative advantage in exporting that product.

References

Ahuja, V., and D. Filmer. 1995. "Educational Attainment in Developing Countries." Policy Research Working Paper 1489. World Bank, Office of the Vice President Development Economics, Washington, D.C.

China Office of the Third Industrial Census. 1996. *Summary of the Third Industrial Census of the People's Republic of China—1995*. Beijing: China Statistical Publishing House.

Coe, D.T., E. Helpman, and A.W. Hoffmaister. 1995. "North-South R&D Spillovers." NBER Working Paper 5048. National Bureau of Economic Research, Cambridge, Mass.

De Long, J.B., and L. Summers. 1993. "How Strongly Do Developing Economies Benefit from Equipment Investment?" *Journal of Monetary Economics* 32(3): 395–415.

Finger, J.M., M. Ingco, and U. Reincke. 1996. "The Uruguay Round: Statistics on Tariff Concessions Given and Received." World Bank, International Economics Department, Washington, D.C.

Grossman, G.M., and E. Helpman. 1993. *Innovation and Growth in the Global Economy.* Cambridge, Mass.: MIT Press.

Hertel, T. 1997. *Global Trade Analysis: Modeling and Applications.* Cambridge: Cambridge University Press.

Hong Kong Monetary Authority. 1997a. "The Economic Integration of Hong Kong and China." Background paper prepared for this report.

———. 1997b. *Hong Kong Annual Digest of Statistics 1996.*

UNCTAD (United Nations Conference on Trade and Development). 1994. *Directory of Import Regimes, Part I: Monitoring Import Regimes.* New York.

———. 1995. *World Investment Report, 1995.* Geneva.

Unirule Institute. 1996. "Measuring the Costs of Protection in China." Paper prepared for the Institute of International Economics. Washington, D.C.

World Bank. 1997a. *China Engaged: Integration with the Global Economy.* Washington, D.C.

———. 1997b. *Private Capital Flows to Developing Countries: The Road to Financial Integration.* New York: Oxford University Press.

chapter eight

Setting the Agenda

China is embarked on an extraordinary voyage of change. Its breathtaking speed and sweep promise new economic horizons and fresh hope for China's huge population. China has telescoped into one generation what other countries took centuries to achieve. In a country whose population exceeds those of Sub-Saharan Africa and Latin America combined, this has been the most remarkable development of our time.

But China is in uncharted waters. No country (let alone one of continental proportions) has tried to accomplish so much in so short a time. China's unique attempt to complete two transitions at once—from a command to a market economy and from a rural to an urban society—is without historical precedent. The complexity of the interactions between these two transitions, the often unpredictable results they bring, the enormous canvas of China itself, and the repercussions for the rest of the world—all make this a task of unparalleled difficulty.

Two questions present themselves. Can China cope? And can the world cope with China? Because so much is at stake, the answer to both questions must be "yes." Eliminating poverty in the most populous nation on earth will be a triumph not only for the Chinese but for all humanity. Ensuring a higher standard of living and a better quality of life for all Chinese will be a victory for development and a force for stability. Ensuring a cleaner environment will be a legacy for future generations and will produce benefits far beyond China's borders. Integrating with the global economy through trade and finance will be for the greater good of China and the world.

China has much to gain from the world, and much to offer it. But fulfilling this potential will not be easy. In China, it will require bold and imaginative leadership, a clear sense of direction, and an unswerving resolve to bring China's two transitions to a successful conclusion. In the world, accommodating China's emerging economic power in a smooth and orderly way will call for vision and statesmanship among the major economic powers, and the active and constructive involvement of international institutions.

Two visions for 2020

The interrelated risks that will challenge China's long march into the twenty-first century give pause for thought. Impressive as China's strengths are, they do not guarantee success. The risks and challenges are strong and varied enough to threaten progress (box 8.1). International experience is littered with examples of economies that have enjoyed prolonged periods of rapid growth only to be followed by setback and stagnation (box 8.2). Conceivably, China might become the first East Asian victim of this phenomenon. It could therefore take either of two very different forms by 2020. Much will depend on the ability and resolve of the authorities to maintain the momentum of reforms.

Two consequences would follow if reforms slow or come to a standstill. First, growth would moderate as the cumulative costs of maintaining increasingly inefficient state enterprises undermine the budget and the banks, burden the rest of the economy, and reduce international competitiveness. And second, the pattern of growth would reflect rising disparities between regions, rural and urban areas, and state and nonstate employment.

This path would lead to "Sinosclerosis," and much of China's promise would have faded by 2020. It would remain a low-income economy with a diminishing presence in world markets. Slow growth and increasing inequality would become a vicious circle, seriously damaging China's efforts to reduce poverty (figure 8.1).[1] Not only would poverty increase in the countryside but it would also grow in urban areas among the elderly living on inadequate pensions, migrants excluded from the urban welfare system and jobs in modern enterprises, and the unemployed unable to find work in a lethargic economy. With growing rural-urban and interprovincial disparities, the prospects for eliminating poverty in poor provinces would recede far into the future (figure 8.2). Cities would become a tinderbox of tensions, with growing numbers of poor living in proximity to an increasingly rich elite that manipulated the laws and systems to its benefit. The gap between coastal and interior provinces would grow, as would differences between city and countryside and between men and women. Foreign investment would be deterred by the opacity of China's trade, legal, and investment systems and by growing instability. Moreover, without evidence of increased openness and transparency in trade, frictions could arise with trading partners and increase the possibility of retaliatory action.

But there is another vision of China in 2020, and it is quite different. This China would be competitive, caring, and confident, having eliminated poverty as it is known today and promising a bright and healthy future for its children. It would be engaged with the rest of the world as an equal and responsible partner in trade and finance, built on modern institutions and the rule of law. It would be a middle-income country with per capita incomes equal to those of Argentina, the Republic of Korea, and Portugal today, enjoying rapid and sustainable growth based on markets and private enterprise.

In this vision, state enterprises would have a crucial role delivering public goods and services, sometimes in competition with private firms. Similarly, state banks would account for a small share of total bank assets with the rest held by dynamic nonstate banks, several of which could be foreign. Improved fiscal management and closer partnership with nonstate entities would produce adequate resources for investments in people, the environment, and infrastructure. China would be cleaner and richer, its people qualified to meet international competition into the twenty-first century. Fluid

BOX 8.1

The risks immediately ahead

This report has highlighted key risks China's economy faces—risks that are complex, immediate, and fundamental. What is more, several of these risks are interrelated and could derail the economy.

- First is the link between banks and state enterprises. As the performance of state enterprises has weakened, so has the financial condition of the state banks. Although the full weight of the government behind the state banks precludes the possibility of a banking collapse, the cost of bailing out the banks is high and rising. The government may have to borrow for this purpose, so its debt service payments could increase. That would mean fewer resources for investing in health, education, infrastructure, and the environment, and for financing the reform of pensions.
- Second is the nexus between state enterprise reforms, labor markets, and inequality. Rising unemployment in some cities may discourage the government from pursuing state enterprise reforms vigorously. Unreformed state enterprises will go on paying excessive wages to an excessively large workforce, supplemented by various social benefits. Again, this will impose a large fiscal burden on society. In addition, grain prices would have to be suppressed to subsidize urban food consumption. This would mean a decline in agricultural incomes relative to urban remuneration. Rural-urban disparities, already the biggest contributor to inequality in China, could widen further. This will stimulate migration into towns and cities, with migrants able to find only low-wage employment in nonstate firms. Together with emerging biases against women, the effect would be to segment labor markets further along regional and gender lines. As inequality grew, China could begin to look less like its East Asian neighbors and more like Latin America.
- Third, delayed enterprise reforms could slow growth and hamper efforts to clean the environment. The slower growth is, the harder it becomes to replace the existing stock of capital with capital stock that embodies the latest environmental standards. And without reforms, enterprises will have little incentive to reduce pollutants. Pollution (of both air and water) could rise sharply, exacerbating mortality rates and the environmental problems of neighboring countries (and perhaps the world).
- Fourth is the government's fiscal position and the system of intergovernmental transfers. Without reform, a broad range of needed public expenditures in health, education, infrastructure, and the environment would be jeopardized, as well as the government's capacity to influence rural-urban and interprovincial income disparities. The same would be true of the other areas of reforms requiring government financial support, such as pensions, banking, and state enterprises.

labor markets would absorb the massive structural changes in the economy, giving firms and workers freedom to negotiate wages and conditions.

By 2020 this China would be the second largest trading nation in the world and a major force in grain and energy markets. Its legal system would protect the rights of foreign as well as domestic owners of physical and intellectual property. In financial markets, integration would bring a larger share of foreign investment and portfolio flows. China would have correspondingly greater weight and voice in international institutions.

These two visions of China in 2020—Sinosclerosis or an agile, modern, and adaptable economy—represent the two extremes on a broad spectrum of possibilities. Where China goes on this spectrum will depend largely on its ability to maintain the momentum of reforms. So far, China's leaders have chosen reform over inaction. We have every reason to think they will continue to do so. But if problems became more complicated, it is not inconceivable that resolve could waver and hesitation set in. If this happens, it is important to step back and reconsider the options. To restore a reformer's determination, there is nothing like staring occasionally into the abyss.

China's agenda to 2020

China's agenda to 2020 should build on the country's strengths—a remarkably high savings rate, a strong record of pragmatic reforms, relative stability, a disciplined and literate labor force, a supportive diaspora, and a growing administrative capacity. These strengths have driven China's growth in the last two decades of this century. They could do the same in the first two decades of the next.

To nurture these strengths and use them to good effect, China's agenda will need to develop along the lines suggested in the earlier chapters. Broadly, they span three comprehensive and interrelated principles. First, to encourage the spread of market forces, especially through reforms of state enterprises, the financial system, and labor markets. Second, to deepen integration with the world economy by lowering import barriers, increasing the transparency and predictability of the trade regime, and gradually integrating with international financial markets. Third, to redirect government toward making markets work better by building the legal, social, physical, and institutional infrastructure of the economy.

BOX 8.2

Three lessons from two giants

Prolonged rapid growth does not necessarily make economies invulnerable. Take the experience of two large economies in Latin America—Brazil and Mexico. Both enjoyed rapid growth between 1950 and 1980, only to stagnate later (see table). Their stagnation was triggered by external factors—the two oil price shocks, falling commodity prices, and the debt crisis of the 1980s. But the vulnerability of these two economies, and their inability to adjust, pointed to internal failings that had been building for several years. In retrospect, the reasons are not hard to find. Governments in both countries:

- Followed strong import substitution policies that initially stimulated growth but eventually became a source of inefficiency, corruption, and rent seeking.
- Expanded public ownership in the economy at the cost of large fiscal deficits, heavy external debt, high inflation, and crippling bad debts within the banking system.
- Administered key prices that exacerbated inequalities, increased regional disparities, and distorted production and consumption.

China's economic policies today bear passing resemblance to those that weakened the Brazilian and Mexican economies. Does this mean that the same fate awaits China? Not necessarily, especially if China maintains its momentum on economic reforms.

Real GDP growth in Brazil and Mexico

(annual percentage change)

	1950–80	1980–95
Brazil	6.8	2.0
Mexico	6.4	1.8

Source: Maddison and others 1992; World Bank staff estimates.

FIGURE 8.1

Growth is important for poverty reduction . . .

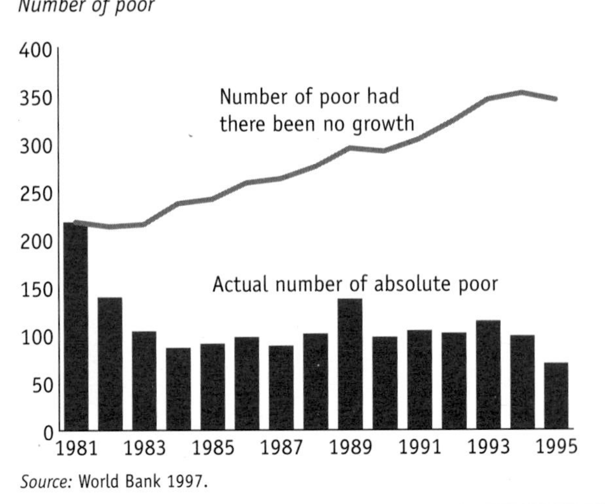

Source: World Bank 1997.

FIGURE 8.2

. . . but growth alone is not enough

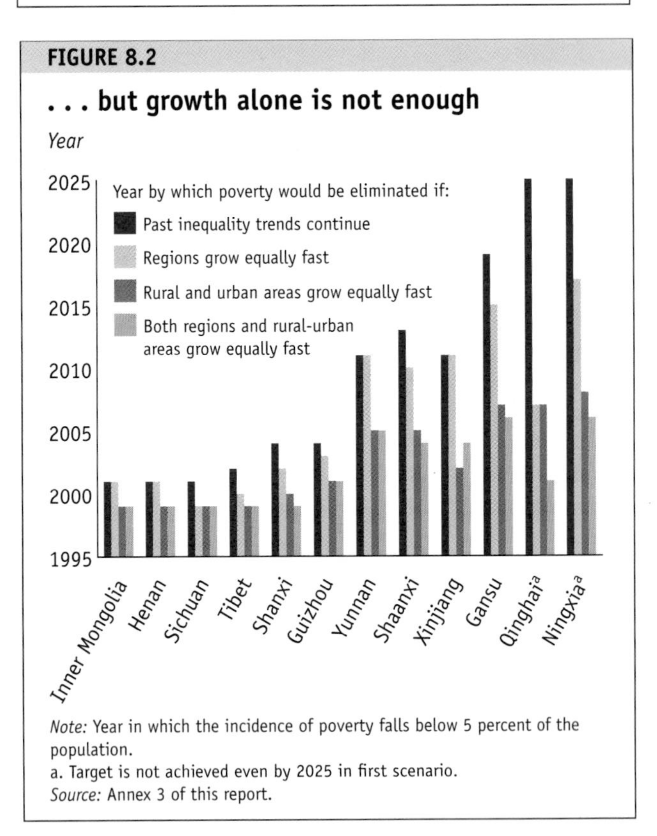

Note: Year in which the incidence of poverty falls below 5 percent of the population.
a. Target is not achieved even by 2025 in first scenario.
Source: Annex 3 of this report.

Extend and deepen markets

Separating state enterprises and commercial banks from government is arguably the most immediate and important of all ongoing reforms in China. The government's decision to focus on developing 1,000 of the largest state enterprises while loosening controls on the remaining 314,000 is a step in the right direction. But much more needs to be done. All enterprises should be fully exposed to domestic and international competition. Just as important, their ownership should be diversified toward households and nonstate institutions. This cannot be emphasized enough. China's future lies with vibrant, competitive, and private firms in industry and services. The earlier success of township and village enterprises and the recent rapid expansion of privately and individually owned firms are evidence of the enormous latent energy of China's dynamic and enterprising people. The government must recognize this potential and design policies to support and shape it.

In the financial sector this means government at all levels should stop interfering in the lending decisions of state commercial banks. Loans to state enterprises need to be subjected to the same scrutiny and incorporate the same safeguards as loans to other borrowers. Banks must begin to build trust, relationships, and information channels with nonstate firms. They must establish systems for management information, risk evaluation, creditworthiness analysis, and provisioning for bad debts. If they manage all this, state banks will be reinventing themselves as commercial institutions with strong portfolios and an adequate capital base. The government could stimulate the transformation by gradually liberalizing interest rates and encouraging competition (initially with domestic banks and eventually with foreign ones too).

In addition, a flourishing capital market is needed to allocate resources, diversify risk, and raise returns to investors. It will be a channel for the government to dilute its holdings in state enterprises and an intermediary for long-term investment, especially in infrastructure. The experience of "red chip" shares traded in Hong Kong shows how successful such a strategy could be.[2] One key to strong capital markets is a strong banking system. Beyond that, the government needs to continue its efforts to strengthen the regulatory oversight of primary and secondary equity markets, develop the fledgling corporate bond market, and install a legal framework for the issuance of asset-backed securities to finance infrastructure projects.

However determined is the reform in state enterprises and the financial sector, it will be less effective if not matched by continued reform of labor markets. In urban areas, delinking social security benefits from employment in state enterprises will separate the commercial decisions of managers from the social concerns of government and free workers to switch jobs without fear of losing their pensions, homes, and access to health and education. More information on employment opportunities and better retraining of laid-off workers will improve the economy's ability to match the demand for labor with supply. And gradually removing restrictions on migration will help the poor find better income-earning opportunities, while protecting the rights of migrants will shield them from possible abuse by unscrupulous employers.

Earlier chapters also described how markets can be deepened in individual subsectors of the economy. In agriculture, for example, there is a need to inject competition into the procurement and distribution of grain. The government could aim to gradually reduce its procurement of total marketed grain from 75 percent today to 25 percent by 2020, leaving room for retail and wholesale markets to develop and for nonstate companies to expand their purchasing and retail networks.

Similarly, the government could meet environmental objectives through greater use of market mechanisms. It could phase in coal taxes to reflect social costs more closely and to encourage conservation and coal-washing technologies. Higher taxes on gasoline and diesel would also help promote fuel economy and protect the environment. Finally, large increases in air and water pollution levies could yield large returns.

It might be tempting for the Chinese to design industrial policies in the way Japan and Korea did in the 1950s and 1960s. It is debatable whether that would be wise; what is certain, however, is that it would be difficult. Freedom would be circumscribed by demanding standards for entering the World Trade Organization and the watchful eye of trading partners. So rather than using protection or subsidies to encourage growth in key subsectors, the government should focus on developing sound competition and technology policies that reward innovation, risk taking, and good management.

Integrate with the world economy

There is no better way of intensifying competition and encouraging the spread of technology than open trade in goods and services. In an era of rapid globalization, with huge advances in technology and the information revolution, the countries that open their doors to fresh ideas and new concepts will be the ones that prosper. Since 96 percent of the world's research and development is undertaken in industrial countries, much technical progress can be captured cheaply by importing their capital goods. Competition and the free flow of economic information can spread the benefits of imported technology quickly throughout the economy, multiplying its initial benefits.

Accelerating this virtuous circle of trade, technological progress, and growth would require a firm timetable for reducing trade restrictions to acceptable international levels. It will need to be supplemented by better trade administration, dispute resolution, and protection of intellectual property rights.

Easier said than done. The authorities recognize that import liberalization will impose short-term costs, as firms will meet import competition by laying off workers and investing in new equipment. Those that do not adjust will have to close. Unfortunately, many of them will be capital-intensive and scale-sensitive—and in regions where unemployment is already high. But the prospect of difficulties in the future should not deter the authorities from acting now. In rapidly growing countries, adjustment tends to be quicker and less disruptive. Moreover, whatever the short-term costs, they will be far outweighed by the long-term benefits of an open economy that derives its strengths from flexibility, competitiveness, and the strong foundation of comparative advantage.

As China proceeds with trade liberalization, the government should not lose sight of integration with the international financial system. Increased portfolio flows would help lower the cost of capital and deepen domestic financial markets. And more foreign direct investment would bring management know-how, the latest capital equipment, and new techniques in international marketing.

Moreover, markets are already integrating China into the world financial system. Large amounts of private capital flow into and out of China, sometimes illegally. The issue for the government is how the forces of integration can be managed for the benefit of the economy and how high growth can avoid being damaged by destabilizing capital flows. Considering all the benefits and the risks of financial integration, it would be wise to proceed cautiously. But procrastination is no answer either. Government inactivity is a sure way to spawn unpredictable and potentially destabilizing movements of capital.

Redirect government to make it more effective

"More market" does not mean less government. It means different government. Milton Friedman once said that China had "both too much government and too little"—too much in production and investment controls and too little in the rule of law, macroeconomic management, and the provision of public goods and services.[3] This has to change. Government energies need to shift away from direct involvement in productive activities and toward two areas in particular: first, more spending on such priority areas as education, health care, agricultural research, infrastructure development, environmental protection, and support for vulnerable groups in society; and second, the development of transparent and participatory institutions that promote the rule of law and a stable economic environment.

Reshaping government expenditures. In the future the government should redirect public resources away from subsidies and investments in state enterprises and toward delivering more and better public goods and services that only it can provide. This report has identified many potential candidates that deserve the government's attention when preparing its future budgets. These include:

- Additional education spending to reduce the burden of school fees borne by the rural poor; to ensure that girls have access to education, especially in rural areas; to meet the government's target of nine mandatory years of schooling by 2000; and to expand university education to support China's need for professional managers and higher-level skills.
- More spending on health, making the rural poor the main beneficiaries, and focusing on immunization, infectious and parasitic diseases, noncommunicable diseases, and the prevention of injuries at work and on roads.
- Substantially more spending on agricultural research and extension to boost grain yields.
- Infrastructure investments in flood control and irrigation, natural gas distribution, public transit, municipal water treatment facilities, and rural roads (especially in interior provinces).
- Protection of the poor and the old, with special emphasis on the welfare of women; for the poor, a renewed stress on basic health and education, combined with assistance to find employment in neighboring, fast-growing provinces; for the elderly, financing the transition from pay as you go pensions to a three-pillared scheme.

The expenditure shortfall in these high-priority areas has been estimated at 4.6 percent of GDP. This is conservative because it does not include extra spending to support future economic reforms—such as redundancy payments for laid-off workers of state enterprises, monetized housing benefits, or the costs of reforming the banking system. Raising revenues to cover these needs will require a huge improvement in tax collection. Even if that is achieved, taxation as a share of GDP will still be below the level in most other countries.

Building institutions. Over time, direct government control of economic decisions must be replaced by legal norms and procedures, and the role of institutions. Simply put, reforms mean that laws and institutions now really matter.

Laws in China have shown the same rapid growth as GDP. As many were scheduled for consideration in the 1993–98 session of the National People's Congress as were enacted between 1949 and 1992. Yet the legal system operates poorly because laws are not yet properly enforced. One step will be to develop the legal profession; another will be to improve legal education.

Corruption is a growing problem in China. The government has moved quickly to contain it, but much remains to be done. Prosecuting the corrupt (including the corrupter) will help. China will also have to reduce incentives for corrupt behavior by lessening the discretionary power of officials, increasing transparency in public finances, and building a competent and honest civil service.

Transparent and responsive institutions will also be invaluable in the preparation and implementation of economic policy. The budget should include many revenues and expenditures that are now extrabudgetary. Exposing official plans to wide public discussion will strike a balance between planners' priorities and people's preferences. The budget, and its record of implementation, should be publicly available. Government agencies should be required to stay within explicit spending limits—and held accountable if they fail.

Also important for markets and growth are the institutions that ensure macroeconomic stability. Of these, the most important is the central bank. Improving its capacity to manage monetary conditions using indirect, market-friendly instruments is central to the economy's future stability. So is its capacity to supervise banks and other financial institutions. As the financial system changes, there is always a danger that risky activity by financial intermediaries will produce systemic shocks. Examples of bank collapses abound in other countries, often resulting in huge costs in forgone economic growth. China has already experienced some isolated problems of this kind and can ill afford repetition on a larger scale as it navigates the treacherous waters of financial sector liberalization.

In summary, China's policy agenda to 2020 calls for further transition and transformation. Naturally, as the economy progresses, the complexities of the issues will increase. But so will the ability of the government to deal with them. There is no single model that China could or should pursue. It has learned from the experience of other countries and adapted policies to suit its unique circumstances. And increasingly, it can look to its own experience to identify policies that promote rapid and sustainable development. Indeed, the essence of market economies—that the pursuit of individual self-interest, combined with open trade and minimalist (but effective) government, can unleash astonishing progress—is embodied in one of China's regions: Hong Kong.

The world and China

The transformation of an economy as large as China's, from low- to middle-income status, from rural to urban, from agricultural to industrial (and services), will inevitably cause ripples in the world economy. By 2020 China could be the world's second largest exporter and importer. Its consumers may have a purchasing power larger than all of Europe's. China's involvement with world financial markets, as a user and supplier of capital, will rival that of most industrial countries.

The emergence of China as a force in world markets is often described as unprecedented. Is it? In fact, the world has seen this happen twice before: the United States in the late nineteenth century, and Japan in the twentieth. If China's trade were to expand as projected in this report, its incremental share of world trade in the twenty-first century would be no different from what those two countries achieved (figure 8.3).

It is revealing that the United States and Japan emerged as trading powers in a world economy with few trade barriers. The second half of the nineteenth century was a halcyon era of open markets that supported the expansion of U.S. trade. The second half of the twentieth century was little different, helping Japan become an industrial giant. During both periods world economic growth was buoyant. Benefits from increased trade were reflected in higher incomes and standards of living.

The emergence of the United States and Japan as trading powers was not merely good for their own economies. It was good for the world. The same should hold true for China. Indeed, the world is better equipped

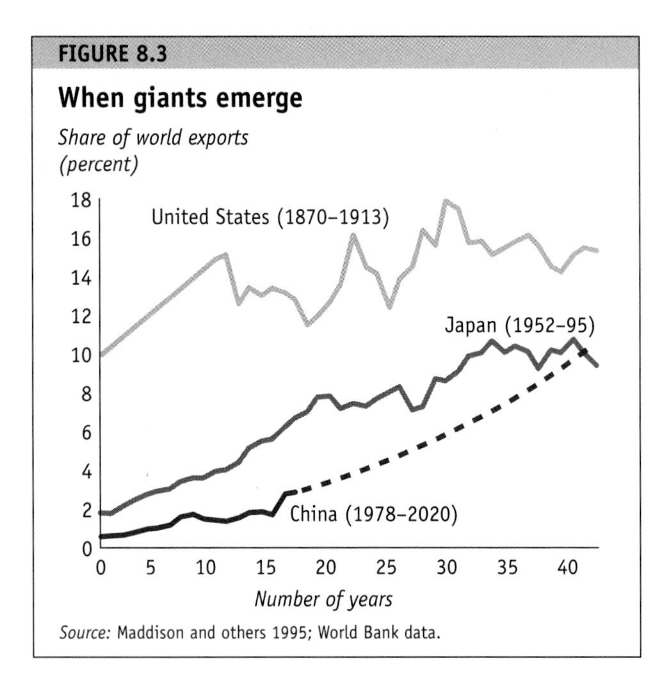

Source: Maddison and others 1995; World Bank data.

to handle the emergence of a major trading nation than it was 50 or 150 years ago. Stronger trading links, more sophisticated financial institutions, and far superior communications make it easier for economies today to adjust to shifts in global conditions. The experience of the two oil shocks in the 1970s showed how open economies coped better than closed ones. Closing markets to China now may hurt its growth prospects, but more important, it will damage other countries' capacity to adjust and benefit from China's dynamism.

Provided markets remain open, China's rapid integration with the global economy will have differing—but mostly favorable—effects on the world. Industrial countries will benefit unequivocally from China's rising demand for capital- and knowledge-intensive manufactures, and through significant terms of trade gains. Only their labor-intensive industries will face a squeeze, but much of that structural change has already occurred.

Among developing countries the effects of China's integration will depend on how much they trade directly with China and how much they compete in third markets. Countries that trade heavily with China (such as Korea) will gain significantly. Low- and middle-income countries that are close competitors with China (such as India and Indonesia) will experience some terms of trade losses, but their total trade will keep growing, and perhaps their world market share of labor-intensive manufactures as well.

In summary, China's rapid growth and liberalization will be an opportunity for the world economy, not a threat. It would be ironic, then, if China were to be denied the same open markets and world conditions that Japan and the United States enjoyed. But trade issues have become more sensitive and complicated, so rapid export growth in the next century may prove more difficult for China. If China's trading partners were to raise barriers to its exports and China were to retaliate, the losses to everybody could be large. To avoid this, China's accession to the World Trade Organization is imperative. Trade relations will become easier to handle, and China's engagement with its trading partners in a multilateral setting will strengthen dialogue and encourage cooperation.

Conclusion

The past two decades have seen sustained, rapid modernization unlike in any other period in China's long history. The next two decades promise more of the same. The huge risks that China faces could yet take the shine off this potential. But with resolute leadership at home and statesmanlike policies from the world's industrial powers, China can overcome these challenges. One-fifth of humanity would then have within its grasp the power to break free of the shackles of poverty and underdevelopment and accomplish what could become the most remarkable economic transformation the world has ever seen.

Notes

1. World Bank (1997).
2. "Red chip" shares traded on the Hong Kong Stock Exchange are the shares of mainland Chinese state enterprises registered in and managed from Hong Kong.
3. Quoted in Rohwer (1996, p. 141).

References

Maddison, A. and others. 1992. *Brazil and Mexico: Political Economy of Poverty, Equity, and Growth.* New York: Oxford University Press.

Maddison, A., and others. 1995. *Monitoring the World Economy, 1820–1992.* Paris: Organisation for Economic Co-operation and Development.

Rohwer, J. 1996. *Asia Rising.* London: Nicholas Brealey Publishing.

World Bank. 1997. *Sharing Rising Incomes: Disparities in China.* Washington, D.C.

annex one

Accounting for China's Growth

This annex describes the growth accounting methodology underlying table 1.1.[1] It first presents the basic growth accounting framework that generates the conclusion that a large share of China's growth cannot be explained by increases in inputs. It then considers the following three questions:

- Does China's large growth residual reflect deficiencies in the data such as those discussed in box 1.1?
- Can differences in technology account for China's large growth residual relative to comparator countries?
- How much of the growth residual can be attributed to the process of structural change discussed in chapter 1?

Preliminaries

The basis of growth accounting is an aggregate value-added production function, $Y = F(A, K, H, L)$, which expresses value added Y as a function of primary inputs, such as physical capital K, human capital H and raw labor L, and an unobserved factor A. Differentiating this production function results in the following expression relating growth in output g_y, to growth in inputs (g_K, g_H, and g_L), and growth in the unobserved factor, g_A:

$$g_y = \beta_K g_K + \beta_H g_H + \beta_L g_L + \beta_A g_A$$

The parameters, β_K, β_H, β_L, and β_A, are known as the "output elasticities" of the various factors of production and measure the percentage increase in output resulting from a percentage increase in inputs. As such, they summarize the production technology available to a country.

Since the first three factors of production (physical capital, human capital, and labor) are in principle observable, the first three terms in the above expression correspond to the portion of growth that can be "explained" by increases in inputs. The fourth term, which consists of the portion of GDP growth accounted for by to increases in the unobserved factor, A, is the unexplained "growth residual." This residual may be interpreted narrowly as productivity growth or more broadly as a "measure of our ignorance."[2]

The basic growth decomposition presented in table 1.1 is based on data from readily available sources on GDP, capital stocks, the labor force, and human capital.[3] The output elasticities of the three factors of production are assumed to be 0.4, 0.3, and 0.3, respectively.[4] As indicated in table 1.1, this decomposition suggests that 46 percent of growth, or 4.3 percentage points of growth per year, is due to factors other than increases in primary inputs. Since this unexplained portion of China's growth is unusually large in comparison with other countries (both as a share of growth and in absolute terms), it is useful to ask how robust this finding is.

Data deficiencies

Data deficiencies cloud the measurement of GDP growth in China. In particular, the unusually slow growth of national accounts deflators for consumption and investment relative to other measures of price increases suggests that real growth in these variables may be overstated. These difficulties also have implications for the measurement of the capital stock, which is measured as the cumulation of past investment flows. If investment is overstated, growth in the capital stock will be overstated as well.[5] The net effect of these data deficiencies on the growth residual is unclear, since output growth may be overstated and growth in capital input may also be overstated.

Table A1.1 summarizes the net effect of the data deficiencies discussed in box 1.1 on the unexplained portion of growth in China. The alternative GDP growth rate is obtained in the same manner as described in box 1.1; the growth rate of the capital stock reflects the alternative, lower investment rate implicit in the GDP growth rate adjustment.[6] The last row reveals that the unexplained portion of growth is essentially unaffected by data concerns, since the decline in the GDP growth rate is offset by lower growth in capital input.

Differences in technology

Estimates of the output elasticities, which summarize the available production technology are very difficult to obtain. One approach is to assume that the production function exhibits constant returns to scale and that perfect competition prevails.[7] In this case, economic theory predicts that the output elasticities are equal to the shares in value-added payments to the factors of production. This presents two obvious difficulties: how to justify the theoretical assumptions and how to obtain reliable data on factor payments, especially wages by skill level (to control for differences in human capital).

An alternative approach is to estimate the output elasticities econometrically.[8] This approach assumes that the fourth term in the growth accounting equation

TABLE A1.1
Output and input growth, 1978–95
(percent per year)

Measure	Official	Alternative
GDP	9.4	8.2
Physical capital	8.8	7.9
Human capital[a]	2.7	2.7
Labor force[a]	2.4	2.4
Unexplained share of growth (percent)	46	43

a. Growth rates calculated over the period 1978–93.
Source: World Bank staff estimates; see note 3 at end of chapter.

can be treated as the residual term in a regression equation, so that the other three output elasticities can be estimated using regression methods. The advantage of this approach is that it requires far fewer restrictions on the form of the production function. However, it raises new difficulties. In particular, the right-hand side variables in such a regression will be correlated with the residual term (since, for example, one would expect the capital stock to increase in response to increases in productivity). Unless suitable instruments can be found, this will result in biased estimates of the output elasticities.[9]

In light of these difficulties, this report takes the more pragmatic approach of simply assuming plausible values for the output elasticities and then examining the extent to which the conclusions depend on reasonable variations in these parameters. In tables 1.1 and A1.1 the output elasticities of physical capital, human capital, and labor were assumed to be 0.4, 0.3, and 0.3, respectively. Here, we briefly examine whether the main conclusion of table 1.1—that China's growth residual is unusually large—depends on this assumption.

Consider, for example, the possibility of increasing returns to scale in the production function (which implies that the output elasticities sum to a value greater than one). Kim and Lau (1994, 1995) have argued that it is important to allow for this possibility because in the presence of increasing returns, increases in inputs lead to more than proportional increases in outputs, so that more of output growth can be "explained" by increases in inputs. In their econometric estimates of production functions Kim and Lau find evidence of substantial increasing returns in the Asian economies but not in the G-5 economies, indicating that the effect is empirically important.

Can increasing returns account for China's large growth residual? A simple way to answer this question is to scale up the three output elasticities by a constant corresponding to the degree of increasing returns. For example, Kim and Lau (1995) find evidence that suggests that the extent of increasing returns is about 1.3 in a sample of other Asian economies (that is, a 1 percent increase in all factors of production leads to a 1.3 percent increase in output). Although allowing for increasing returns reduces the growth residual significantly—from nearly half to roughly one-third of growth—the basic conclusion that there is a substantial unexplained portion of growth remains unchanged (table A1.2). Moreover, it is clear that increasing returns cannot account for the difference in the unexplained portion of growth between China and, for example, the Republic of Korea, unless one is willing to argue that increasing returns are much stronger in China than in Korea.

TABLE A1.2
Can increasing returns account for the growth residual?
(percent)

	Unexplained share of growth	
Country/period	Constant returns	Increasing returns (1.3)
China 1978–95	46	30
Japan 1960–93	30	9
Republic of Korea 1960–93	21	-2

Source: World Bank staff estimates.

Structural change and the growth residual

China's two transitions can explain some of its large growth residual. The process of transition has been a potent source of growth as factors, especially labor, have been transferred from sectors in which their marginal productivity was relatively low (agriculture, state-owned industry) to sectors in which their marginal productivity is high (industry and services, nonstate industry).

Estimates of the growth contribution of factor reallocation are presented in table 1.1. According to World Bank estimates, the reallocation of labor out of the state-owned sector contributed roughly 0.5 percentage points to GDP growth between 1985 and 1994, while the transfer of labor out of agriculture over the same period contributed an additional 1.0 percentage point to growth.[10] Similar calculations reveal that about 0.3 percentage points of growth in Japan and the Republic of Korea between 1960 and 1993 can be attributed to the reallocation of labor out of agriculture in these countries. In the United States, labor reallocation across broad sectors since 1960 contributed little to growth. Data on the sectoral composition of GDP in the United States in the last century are not available, preventing the calculation of the benefits of labor reallocation over this period. However, it is reasonable to assume that factor reallocation played an important role in growth in the U.S. economy over this period of rapid structural change.

Notes

1. This annex draws substantially on Kraay (1996).

2. Although conventional, this decomposition is somewhat arbitrary, as it does not take into account the fact that at least some of the growth in factor inputs is in response to the productivity growth captured by the residual. For a discussion, see Barro and Sala-i-Martin (1995).

3. Data for the comparator countries in table 1.1 were obtained as follows. For Japan and the Republic of Korea, data on physical capital, human capital (measured as average number of years of education of the working-age population), and the working-age population are drawn from Nehru and Dhareshwar (1993). and Nehru, Swanson, and Dubey (1995). Data on the United States are from Maddison (1995).

4. This assumption may be justified based on the empirical work of Mankiw, Romer, and Weil (1992). As discussed in Kraay (1996), there is also a rough correspondence between these output elasticities and those obtained by Kim and Lau (1994, 1995), and Young (1995) using very different methods.

5. In addition to the flow of investment, the growth rate of the capital stock depends crucially on two other factors: the rate of depreciation and the initial value of the capital stock. In comparative studies it is standard practice to assume that the rate of depreciation is constant and equal across countries, so that variation in capital stock growth rates is not tainted by variations in this difficult-to-measure variable. The initial value of the capital stock is more difficult—and also more important—to determine, since a given flow of investment results in a larger capital stock growth rate the smaller is the initial capital stock. However, as discussed in Kraay (1996), China's capital-output ratio in 1978 is comparable to that of the Soviet Union in 1939, after both countries had pursued twenty years of Soviet-style industrialization. This suggests that the initial capital stock figures for China are of the right order of magnitude.

6. Alternative measures of GDP growth are as described in box 1.1. The revised investment figures are cumulated assuming a depreciation rate of 4 percent a year to arrive at revised capital stock growth rates.

7. See Zuliu Hu and Mohsin Khan (1996) and Young (1996) for applications of this methodology to China.

8. See Kim and Lau (1994, 1995) for applications of this methodology.

9. See Burnside (1996) for examples of the fragility of such production function regressions when the instrument sets are varied.

11. See World Bank (1996). Although the analysis in this report covers only the period from 1986 to 1994, it is not unreasonable to extrapolate this over the entire period since 1978. Although the period from 1978 to 1986 saw less labor reallocation across ownership forms, there was more reallocation of labor out of agriculture during this period.

References

Barro, Robert, and Xavier Sala-i-Martin. 1995. *Economic Growth*. New York: McGraw Hill

Burnside, Craig. 1996. "Production Function Regressions, Returns to Scale and External Effects." *Journal of Monetary Economics* 3(2): 177–201.

Hu, Zuliu, and Mohsin Khan. 1996. "Why is China Growing So Fast?" International Monetary Fund, Washington, D.C.

Kim, Jong-il, and Lawrence Lau. 1994. "The Sources of Economic Growth of the East Asian Newly Industrialized Countries." *Journal of the Japanese and International Economies* 8(3): 235–271.

———. 1995. "The Role of Human Capital in the Economic Growth of the East Asian Newly Industrialized Countries." *Asia-Pacific Economic Review* 1(3): 3–22.

Kraay, A. 1996. "A Resilient Residual: Accounting for China's Growth Performance in Light of the Asian Miracle." World Bank Policy Research Department. Washington, D.C.

Maddison, A. 1995. *Monitoring the World Economy.* Paris: Organisation for Economic Co-operation and Development.

Mankiw, N. G., David Romer, and David Weil. 1992. "A Contribution to the Empirics of Economic Growth." *Quarterly Journal of Economics* 107(2): 407–37.

Nehru, V., and A. M. Dhareshwar. 1993. "A New Database on Physical Capital Stock: Sources, Methodology and Results." *Revista de Analisis Economico* 8(1): 37–59.

Nehru, V., Eric Swanson, and Ashutosh Dubey. 1995. "A New Database on Human Capital Stock in Developing and Industrial Countries: Sources, Methodology and Results." *Journal of Development Economics* 46(2): 379–401.

World Bank. 1996. *The Chinese Economy: Fighting Inflation, Deepening Reforms.* Washington, D.C.

Young, Alwyn. 1995 "The Tyranny of Numbers: Confronting the Statistical Realities of the East Asian Growth Experience." *Quarterly Journal of Economics* 100(3): 605–640.

———. 1996. *The Razor's Edge: Distortions, Incremental Reform and the Theory of the Second Best in the People's Republic of China.* Boston: Boston University.

annex two

Modeling Growth and Structural Change

This annex describes the economic model developed for this report and used in chapter 2 to lay out China's long-term growth possibilities. The objective of the model is to provide as simple a framework as possible with which to highlight the changing role of the fundamental determinants of growth in China: productivity growth, savings, and structural change. The next section provides a nontechnical overview, followed by a more detailed presentation of the equations of the model. The following section discusses how the model's parameters are obtained. The final section presents some additional results that supplement the discussion in the text.

Nontechnical overview

The model presented here is a simple multisector version of the Solow growth model. The Solow model consists of the following essential elements:

- A production function that relates output to primary inputs of capital and labor
- A consumption-savings rule that determines how much of output is available for investment
- An accumulation equation that describes the growth of the capital stock as a function of investment and depreciation.

With these three elements in hand, it is possible to describe the trajectory of output growth given an initial point.

In the long run there are only two sources of growth in the Solow model: growth in the labor force and labor-augmenting technical progress.[1] Along the transition to this long-run growth rate, however, GDP growth is initially higher than its long-run value; it declines over time as diminishing returns set in with the accumulation of capital. During the transition the savings rate emerges as an additional factor influencing the growth rate. Holding constant initial conditions, the higher the savings rate, the higher the eventual long-run level of per capita GDP and the faster the economy grows along the transition path in order to reach its long-run level of development.

The model presented here shares these features of the basic Solow model and is complicated only by the assumption of a slightly more elaborate production structure. Instead of a single good, there are N sectors in the economy, each producing a different good. The N goods are produced using primary factors of land, labor and capital, and intermediates. These goods in turn are used as intermediates in the production of other goods, for final consumption, and for investment.

In the model it is useful to think of the economy as populated by households and firms. Firms demand factors of production, which they use to produce the N goods available in the economy. Households supply labor to firms and receive wages and the profits of firms (value added) as income. An exogenously given fraction of this income is consumed, and the remainder is used to purchase goods that are turned into capital goods used for investment purposes.[2]

In each period the following events occur. Given input and output prices and given their capital stock and level of productivity inherited from the previous period, firms demand labor and intermediates in order to produce output. Households supply their labor, demand goods for consumption, and turn their savings over to a capital goods sector, which uses them to purchase goods used to construct capital goods for investment purposes. Prices adjust to clear the market for goods and labor. At the end of the period firms in each sector purchase capital goods, which they use to increase their capital stocks. The labor force and technology increase exogenously, and the next period begins.

In order to provide a more realistic picture of the Chinese economy, frictions are introduced into the model that prevent labor and capital from being optimally allocated across sectors. This is meant to capture, in a simple way, the effects of real-world barriers to labor mobility, which have kept a substantial portion of the agricultural labor force underemployed, and distortions in the investment system, which bias investment toward industry at the expense of agriculture and services. As these frictions diminish over time, improvements in the efficiency of factor allocation provides an additional, albeit modest, source of growth.

The structure of the model was deliberately kept very simple in order to focus on growth fundamentals, such as savings, productivity growth, and factor reallocation. Keeping the model simple necessarily meant that important factors, including trade and government, had to be ignored. By assuming that the economy is closed, the model cannot capture the (poorly understood) effect of trade on growth or the contribution of foreign savings to domestic capital accumulation. The effects of China's growth on world trade are analyzed in a separate, complementary exercise (chapter 7). The model also ignores the fact that government has a crucial role to play in shaping the environment in which households and firms operate. Since modeling these complex interactions in a convincing manner is difficult, these issues are addressed outside of the formal model.

The model

At the beginning of each period the representative firm in sector j produces gross output (Q_j) using a constant elasticity of substitution (CES) production technology of the form:

$$Q_j = \left[(\beta_{Fj})\left(F_j^{\frac{\mu-1}{\mu}}\right) + (\beta_{Kj})\left(K_j^{\frac{\mu-1}{\mu}}\right) + (\beta_{Lj})\,(A_jL_j)^{\frac{\mu-1}{\mu}} + (\beta_{Mj})\left(M_j^{\frac{\mu-1}{\mu}}\right)\right]^{\frac{\mu}{\mu-1}}$$

where $j = 1, ..., N$ indexes the sectors of the economy. F_j is a fixed factor of production, K_j is the capital stock inherited from the previous period, L_j is labor input, and M_j is a CES aggregate of the intermediate goods (that is,

$$M_j = \left[\sum_{i=1}^{N}\phi_{ij}\left(X_{ij}^{\frac{\varepsilon-1}{\varepsilon}}\right)\right]^{\frac{\varepsilon}{\varepsilon-1}},$$

where X_{ij} is the intermediate use of good i by a firm in sector j). The elasticity of substitution between F_j, K_j, L_j, and M_j is given by μ, while ε is the elasticity of substitution between intermediates. The parameters β_{Fj}, β_{Kj}, β_{Lj}, β_{Mj}, and ϕ_{ij} vary across sectors. A_j represents the level of the labor-augmenting state of technology, and the growth in A_j corresponds to productivity growth. Given K_j and A_j, at the beginning of each period the firm demands labor and intermediates in order to maximize profits. Optimal behavior by firms results in a labor demand function and set of N intermediate demands for each of the N firms in the economy.

In the absence of frictions, optimal firm behavior implies that labor and intermediates will be efficiently allocated across sectors. Since this is an unrealistic approximation for China today, we introduce frictions into the labor market by multiplying each firm's labor demand by a scalar that is greater than one in sectors in which more labor than what would be optimal according to the model is observed and smaller than one in sectors in which less labor than would be optimal is observed. This scalar creates a wedge between marginal products of labor across sectors. As discussed in more detail below in the section on parameterization, the initial values of these scalars were chosen to match the observed distribution of employment in China in the base period; these differentials are allowed to erode exogenously over time.

The representative household allocates its consumption over the N goods available in the economy so as to maximize

$$C = \left[\sum_{i=1}^{N}\alpha_i\, C_i^{\frac{\sigma-1}{\sigma}}\right]^{\frac{\sigma}{\sigma-1}},$$

subject to the constraint that total consumption expenditure is equal to $(1-s)Y$, where s is the savings rate and Y is household income. Since households receive labor income as well as the profits of firms, Y is equal to value added. The elasticity of substitution between goods in consumption is σ. Optimal behavior by households results in a set of N demands for goods for consumption purposes.

The representative firm in the capital goods sector uses households' savings to purchase goods from the N sectors of the economy, I_j^D, in order to produce a homogeneous capital good using the following CES technology

$$I = B\left[\sum_{i=1}^{N}\lambda_i\, I_i^{D\frac{\eta-1}{\eta}}\right]^{\frac{\eta}{\eta-1}}.$$

The elasticity of substitution between investment goods at the sectoral level is η. Optimal behavior by the capital goods firm results in additional N demands for goods for investment purposes.

Markets clear in two stages, as discussed above. First, goods prices adjust to clear the goods and labor markets (the wage is the numeraire). Markets clear when the sum of the N labor demands is equal to the exogenously given labor supply and the sum of the intermediate, final consumption, and investment use demands for each of the N goods in the economy is equal to the corresponding supply. Dropping one of the redundant market-clearing conditions yields a nonlinear system of N equations in N prices that can be solved numerically for the market-clearing price vector.[3]

Next, firms demand investment goods subject to convex adjustment costs and given the price of capital goods. Optimal behavior by firms results in a set of N first-order conditions equating the value of the marginal product of an additional unit of capital with the price of capital goods:

$$p_iQ_i^{\frac{1}{\mu}}\,\beta_{Ki}\left[(1-\delta_i)\,K_i + I_i\right]^{\frac{1}{\mu}} - g'(I_i/K_i) = p_K,$$

where δ is depreciation of capital and $g_i(I_i/K_i) = a_i(I_i/K_i)^2$ is the adjustment cost function. These N equations can also be solved numerically for the optimal distribution of investment. Finally, capital stocks accumulate according to the investment decisions of firms, productivity and the labor force grow exogenously, and the next period begins.

Parameterization

In order to implement the model described above, four types of information are required:

- Values for all the parameters of the production functions and household preferences
- Information on the initial allocation of factors of production in a particular base year
- Values for the parameters that capture the frictions in the labor and capital market
- Reasonable assumptions on the path of savings rates and productivity growth.

These requirements are taken up in turn below.

In the model, there are N=3 sectors: agriculture, industry (including construction), and services. The parameters describing the production side of the model are derived directly from the 1992 input-output (IO) table for China in the following manner. First, the 1992 IO table is aggregated into agriculture, industry, and services. Next, under the assumption of constant returns to scale in production and a unit elasticity of substitution between factors and between intermediates, the output elasticities of labor, capital, and intermediate goods are given by the parameters β_L, β_K, and β_M, while ϕ_{ij} is the elasticity of the intermediate goods aggregate in sector j with respect to intermediate good i. Under the further assumption of perfect competition, these parameters can be identified by the corresponding shares of factor payments in value added, which can be obtained directly from the IO table. In particular,

$$\beta_{L_j} = \frac{w\,L_j}{p_j\,Q_j},\ \beta_{M_j} = \frac{p_{Mj}\,M_j}{p_j\,Q_j},\ \text{and}\ \phi_{ij} = \frac{p_i\,X_{ij}}{\beta_{Mj}\,p_j Q_j}.$$

Since there are constant returns to scale, $\beta_{Kj} = 1-\beta_{Lj}-\beta_{Mj}$.

To complete the parameterization of the production side of the economy, values for the initial level of the productivity parameters, A_j, are needed. Since there is a measure of gross output for each sector (Q_j) and all the other parameters of the production function are available, given information on the inputs of capital, labor, and intermediates in each sector, the A_js can be computed as residuals. Estimates of the labor force and its sectoral allocation are readily available, and since the wage is the numeraire, there is a measure of real labor input for each sector. In the case of intermediates, no information on the allocation of intermediates across sectors other than the IO table is available. Accordingly, the actual values of intermediate uses reported in the IO table was used as a measure of intermediates, valued at 1992 prices. For capital, under the assumption that the returns to capital are equalized, the allocation of capital across sectors is the same as the distribution of factor payments to capital across sectors. Fixing the level of the capital stock in 1992 at 6,000 billion 1992 yuan, this yields a measure of capital input for each sector.[4]

The assumptions that permit the parameters of the production side of the economy to be identified cannot be taken as literally characterizing the Chinese economy in 1992. However, the parameters obtained are roughly in line with the available microeconomic evidence. Moreover, the broad picture that emerges from the projections is relatively insensitive to minor variations in the parameters obtained.

To parameterize the consumption side of the economy, a unit elasticity of substitution in consumption is assumed, and the α_js are identified as the expenditure shares from the appropriate consumption column of the IO table. This yields expenditure shares of .25 on agriculture, .35 on industry, and .40 on services. Since tastes are likely to shift toward services over the course of development, the expenditure share on agriculture is assumed to decline to .20, while that of services is assumed to increase by the same amount over the twenty-five-year projection period.[5]

Finally, the capital goods technology is also assumed to have a unit elasticity of substitution, and the corresponding elements of the investment column of the IO table are used to determine the expenditure shares for investment. Using the appropriate entries from the investment column of the IO table as inputs and total investment, the scale parameter, B, of the investment technology is obtained as a residual.

Table A2.1 summarizes the parameters of the model. Overall the parameters obtained in this manner appear to be plausible and in line with available micro evidence. In a few cases, as indicated in the table, the parameters obtained from the IO table were modified slightly so that they were more consistent with our priors.

Next we turn to the problem of obtaining parameters that describe the rigidities that prevent labor and capital from being optimally allocated across sectors. For the labor market, the observed distribution of employment in 1992 and the production functions estimated from the IO table were used to compute differences in the marginal product of labor across sectors. The labor demands of firms were then scaled by constants reflecting these differences.

TABLE A2.1

Model parameters

	Agriculture (j=1)	Industry (j=2)	Services (j=3)
Output elasticities of:			
Capital (β_{Kj})	.20[a]	.20	.27[b]
Labor (β_{Lj})	.34[a]	.09	.24[b]
Intermediates (β_{Mj})	.36	.71	.49
Fixed factor (land) (β_{Fj})	.1[a]	0	0
Intermediate elasticity of:			
Agriculture (ϕ_{1j})	.39	.10	.30
Industry (ϕ_{2j})	.44	.72	.56
Services (ϕ_{3j})	.17	.19	.42
Investment shares (λ_j)	.04	.88	.08
Consumption shares (α_j)[c]	.25	.35	.40

a. IO table yields values of β_{K1}=.10 and β_{L1}=.54 and provides no information on β_{F1}.
b. IO table yields values of β_{K3}=.31 and β_{L3}=.20.
c. α_1 declines to .20 over projection period, while α_3 increases to .45.
Source: World Bank staff estimates.

TABLE A2.2

Parameterizing frictions in labor and capital markets

	Agriculture (j=1)	Industry (j=2)	Services (j=3)
Labor market frictions			
MPL scaled by	1.92	.59	.50
Implied MPL relative to agriculture	1.00	3.27	3.87
Capital market frictions			
Parameter *a* in $g(I/K)=a(I/K)^2$	400	0	300
Implied MPK relative to agriculture	1.00	0.33	0.37

Source: World Bank staff estimates.

The same procedure cannot be used to estimate differences in the marginal product of capital, because the distribution of the capital stock was obtained from the IO table, rather than from an independent source. As a result, the marginal products of capital are equalized across sectors by construction. Since some information on the distribution of investment in 1992 is available, the initial values of the parameters of the adjustment cost function were chosen so that the model approximately matches the observed distribution of investment. As a result, although the initial capital stock is optimally allocated across sectors, increments to the capital stock are not optimally allocated in the sense that their pre-adjustment cost marginal productivities differ across sectors.

These calculations are summarized in table A2.2. The first row shows the actual values of the scalars multiplying the marginal products of labor; the second row shows the implied differences in marginal products of labor across sectors. The third and fourth rows show the adjustment costs in the capital goods market. The third row shows the actual value of the parameter of the adjustment cost function; the fourth row shows the implied marginal products of capital relative to agriculture.

The final input into the model is a set of assumptions on the future path of the savings rate and of productivity growth. The savings rate is assumed to decline linearly from 40 percent of GDP in 1995 to its long-run value of 35 percent in 2005 and to remain there for the remainder of the projection period. Labor-augmenting productivity growth declines from 5 percent in 1995 to its long-run value of 3.75 percent by 2000 and remains at this value until 2020. Since the production function is Cobb-Douglas given our unit elasticity assumption and since the aggregate labor share in value added is about 0.4, total factor productivity growth is 1.5 percent between 2000 and 2020. Given China's experience with savings rates and productivity growth (see chapter 1), these twc assumptions represent fairly conservative extrapolations of past trends. Variations in these assumptions, of course, lead to large variations in the growth rate over the next twenty-five years, as indicated in table 2.1.

Results

The model's projections of long-run growth and structural change are presented in the second half of chapter 2. Here we briefly supplement these results with a discussion of the contribution of factor reallocation to growth as distortions in the labor and capital market decline in importance over the next twenty-five years.

As discussed in chapter 1 and annex 1, the process of structural change has contributed to growth in China since 1978, since it has resulted in the reallocation of factors out of sectors in which their marginal productivity was low into sectors in which productivity was higher. Simple back-of-the-envelope calculations suggest that labor reallocation alone has contributed 1–1.5 percentage points of growth between 1985 and 1994.

Figure A2.1 summarizes the contribution of reallocation effects to growth over the projection period, decomposing the total factor reallocation effect into a labor and a capital reallocation effect. Labor realloca-

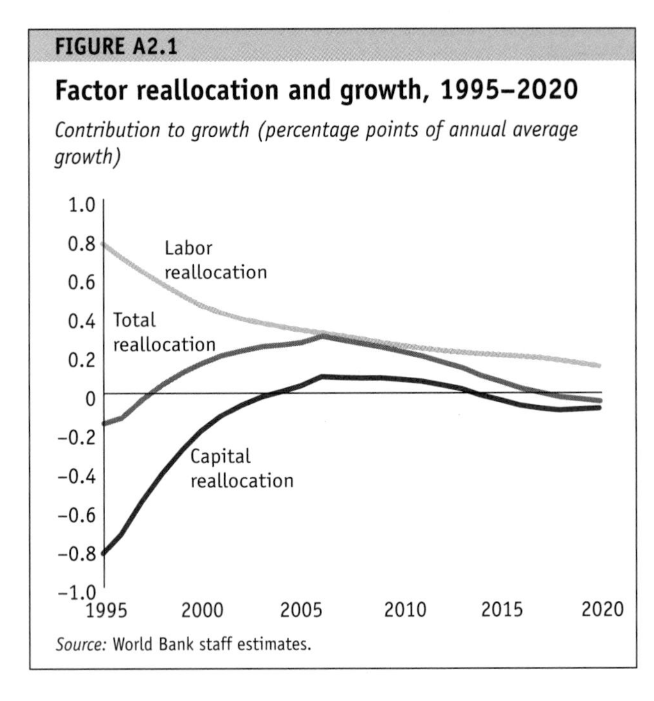

Source: World Bank staff estimates.

tion continues to provide a modest boost to growth of slightly more than 0.5 percentage points initially. However, this contribution to growth dwindles as the barriers to the efficient allocation of labor diminish over time.

More interestingly, capital reallocation effects initially provide a substantial negative contribution to growth, because the barriers to investment (modeled as adjustment costs) result in incremental units of capital being inefficiently allocated to industry rather than to agriculture or services, where their marginal productivity is higher. Eventually, this negative contribution to growth also dwindles as barriers to the efficient allocation of investment decline over time.

Although overall reallocation effects are fairly small (ranging from −0.2 to 0.3 percentage points of annual growth), they point to the importance of a policy environment that facilitates the efficient allocation of factors of production. In particular, they suggest that there are substantial costs in terms of forgone growth associated with policies that bias the investment system toward industry.

Notes

1. In economic jargon the very long run is referred to as the "steady-state" of the model, in which the capital stock per effective worker is constant. For this steady state to exist, technical progress must be labor augmenting.

2. Three key assumptions here keep the structure of the model simple, albeit at the cost of some realism. First, the labor supply is assumed to be exogenously determined, and hence the model cannot account for the cyclical variations in employment that lie at the heart of economic fluctuations. However, since the model is concerned only with China's long-term growth possibilities, the omission is not an important one. Second, savings are assumed to be exogenous, which prevents the (poorly understood) feedback from growth to savings from being modeled. This structure does, however, capture the effect of savings on growth, as described in the text, and, moreover, has the advantage of greatly simplifying the solution of the model. Third, firms are assumed to be purely passive agents, so that the savings/investment decisions of firms are identical to those of households: that is, households "pierce the corporate veil."

3. The model is programmed in Gauss, a matrix programming language for personal computers.

4. Both the level and the distribution of the initial capital stock are important ingredients for the model. The level of the capital stock matters because it determines how far China is from its steady state. The estimate of 6,000 billion yuan is based on Nehru and Dhareshwar (1993). Chow (1993) provides independent estimates of the sectoral distribution of the capital stock, albeit only through 1985. Surprisingly, Chow finds a distribution of the capital stock across the three sectors of 9 percent for agriculture, 62 percent for industry, and 29 percent for services. The distribution implied by the IO table is similar, at 6 percent, 58 percent, and 36 percent.

5. Because of the unit elasticity of substitution in production and consumption, these Engel effects have minimal effects on the pattern of production measured in constant prices.

References

Chow, G. 1993. "Capital Formation and Economic Growth in China." *Quarterly Journal of Economics* 108(3): 809–42.

Nehru, V., Eric Swanson, and Ashutosh Dubey. 1995. "A New Database on Human Capital Stock in Developing and Industrial Countries: Sources, Methodology and Results." *Journal of Development Economics* 46(2): 379–401.

annex three

Projecting Poverty and Inequality

This annex describes the methodology employed in chapter 8 to project poverty and inequality. The projection framework is very simple and embodies no behavioral relationships. Rather, it consists of a set of assumptions on population and income growth that yield projections of average per capita income in rural and urban areas for China's thirty provinces, and a set of assumptions on the future distribution of per capita income in each region. Given these assumptions it is possible to estimate the number of people falling below a specified poverty line in each region in order to arrive at a national projected poverty headcount, and to construct an estimate of overall income inequality. Although the framework is rudimentary, it dramatizes the importance of reducing regional and rural-urban growth disparities if China is to achieve its target of eradicating poverty and stemming increases in inequality.

The next section introduces notation and describes the structure of the assumptions. The last section presents four alternative scenarios.

Counting the poor

The first step in projecting regional poverty is to obtain projections of population and per capita incomes in rural and urban areas in China's thirty provinces. Define

N_j = population of province j
y_j = per capita income in province j

Assume that overall population (per capita income) grows at the rate of g_N (g_y). Let r_{Nj} (r_{yj}) denote population (per capita income) growth *relative to overall growth of that variable* in province j (j=1,...,30). That is, provincial population (per capita income) growth is $g_{Nj} = g_N r_{Nj}$ ($g_{yj} = g_y(r_{yj})$.

To separate urban and rural projected provincial populations and per capita incomes, define the overall share of the population living in urban areas as u^*, and let r_{uj} denote the urbanization rate of province j *relative to overall urbanization.* The projected urban (rural) populations of province j, N_{uj}^* (N_{rj}^*) are then given by

$$N_{uj}^* = u^* \, r_{uj} \, N_j$$
$$N_{rj}^* = (1 - u^* \, r_{uj}) \, N_j$$

Let d_j^* denote future urban/rural relative incomes in province j. Then, predicted urban (rural) per capita incomes y_{uj}^* (y_{rj}^*) are given by

$$y_{uj}^* = y_j^* \, N_j^*/(N_{uj} + N_{rj}/d_j^*)$$
$$y_{rj}^* = y_j^* \, N_j^*/(N_{uj} \, d_j^* + N_{rj})$$

Assumptions on the distribution of income around these average levels are required to arrive at the number of people living below a specified poverty line P. The distribution of income in each region is assumed to be lognormal and the mean log deviation of per capita incomes is assumed to be t_{ij}^*, for $i=u,r$ and j=1,..,30 provinces. Hence, the number of poor in region ij is given by

$$NPoor_{ij}^* = N_{ij}^* \, F(P,\mu_{ij},\sigma_{ij})$$

where $F(P,\mu,\sigma)$ is the cumulative lognormal distribution distribution function with parameters

$$\sigma_{ij}^2 = 2 \, t_{ij}^*$$
$$\mu_{ij} = \ln(y_{ij}^*) - (1/2) \, \sigma_{ij}^2$$

Summing over all provinces yields an estimate of the number of people living below the poverty line in the specified year.

These projections can also be used to construct an estimate of future income inequality. First, the identity that relates the national mean log deviation of per capita incomes (t^*) to the projected regional average per capita incomes and regional mean log deviations is used:

$$t = \sum_i p_i \ln(\tfrac{y}{y_i}) + \sum_i \sum_j p_{ij} \ln(\tfrac{y_i}{y_{ij}}) + \sum_i \sum_j p_{ij} \, t_{ij}$$

where $i=u,r$, j=1,...,30 provinces, y is the national average per capita income, y_U (y_R) are urban (rural) average per capita incomes, p_U (p_R) are the shares of the population living in urban (rural) areas, p_{ij} is the share of the population living in region ij, and t_{ij} is the mean log deviation of per capita incomes in region ij. The resulting national mean log deviation of per capita incomes is then translated into a Gini coefficient that can be used in international comparisons.[1] Clearly, the estimates of income inequality arrived at in this manner will represent upper bounds, since the projection framework does not capture the effects that interprovincial and rural-urban migration will have on regional income differentials.

Scenarios for reducing poverty

The key components of the above projection framework are the initial values for population, per capita income, and the distribution of income in rural and urban areas in China's thirty provinces and the assumptions that yield the future values of these variables. The base year for the projections is 1995, the most recent year for which comprehensive data on rural and urban per capita income for China's thirty provinces are available. In all scenarios we maintain the following assumptions.

- National population grows at an average rate of 1 percent a year and real per capita income grows at an average rate of 5.5 percent a year, consistent with the projections in chapter 2.

• Relative to national population growth provincial population growth rates remain constant and equal to their historical values over the period 1978–95.
• The urban share of the population grows at an average rate of 1.5 percent a year.
• Rural and urban income inequality in each province, as summarized by the the mean log deviation of per capita incomes, remains unchanged from its current level.[2]
• The poverty line in 1995 yuan is 561 yuan.[3]

Given these basic assumptions, we consider four alternative scenarios.

• *Past trends continue.* Provincial relative per capita income growth rates are assumed to remain at their historical values over the 1978–95 period. Urban income growth is assumed to exceed rural income growth by 3.5 percent, as it has over the 1981–95 period (that is, d_i^* grows at 3.5 percent a year from its 1995 base level).
• *Regional growth disparities are eliminated.* Provincial per capita incomes are assumed to grow at the same rate (r_{yj}=1 for all provinces). Urban-rural relative income growth rates are assumed to continue to differ, as in the first scenario.
• *Urban-rural growth disparities are eliminated.* Urban-rural relative incomes are assumed to remain at their 1995 levels (that is, d_i^* is constant and equal to its value in 1995), but regional growth disparities are assumed to persist at their historical levels.
• *Regional and urban-rural growth disparities are eliminated.* All provinces are assumed to grow at the same rate, and urban-rural relative incomes are assumed to remain at their current levels.

For each scenario the year in which the poverty rate in each province $[(NPoor_{iR}^* + NPoor_{iU}^*)/N_i^*]$ falls below 5 percent of the population is calculated. These figures are plotted in figure 8.2. for the twelve provinces in which the incidence of poverty exceeds 5 percent of the population today. The 5 percent threshold is used because the key element in these projections is the rate at which the area in the tail of the lognormal distribution declines as the mean of the distribution shifts upward. This elasticity becomes very large as the area in the tail of the distribution becomes small, rendering the projected poverty rate at low levels very sensitive to changes in mean income. The somewhat arbitrary threshold of 5 percent is used to minimize this source of sensitivity.

In addition, the implied Gini coefficient associated with each of these scenarios is computed, at five year intervals (figure A3.1). Although by assumption within-group inequality remains unchanged, overall inequality changes markedly as regional and rural-urban per capita incomes diverge under the various scenarios.

FIGURE A3.1

Growing apart? Projected income inequality under alternative scenarios

Gini coefficient

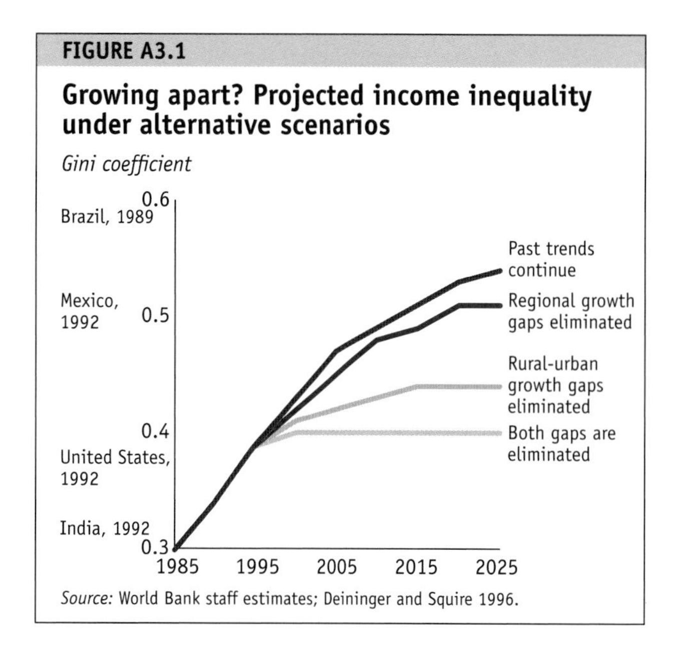

Source: World Bank staff estimates; Deininger and Squire 1996.

Notes

1. This translation is facilitated by the additional assumption that the aggregate distribution of per capita incomes is lognormal.
2. The most recent year for which a full set of rural and urban income distributions by province is available is 1992 (see World Bank 1997). We assume that the mean log deviation of rural and urban incomes in each province in 1995 is equal to its level in 1992, scaled up by a factor of 1.15 to capture the growth of inequality due to factors other than rural-urban and regional growth disparities over the period 1992–95. The factor of 1.15 was chosen to ensure that the overall poverty level in 1995 obtained by summing the rural and urban poverty headcounts in each province was roughly equal to poverty figures based on the available national poverty figures. Evidently, the assumption that within-group inequality does not change is optimistic in light of past trends. Yet, as discussed in chapters 1 and 4, the main source of the increase in overall income inequality has been differences in provincial, and especially rural-urban income, growth rates.
3. This figure is based on the Chinese absolute poverty line of 318 (1990) yuan, extrapolated to 1995 using the rural CPI. See World Bank (1992).
4. Deininger and Squire (1996).

References

Deininger, K. and L. Squire. 1996. "A New Data Set Measuring Income Inequality." *The World Bank Economic Review* 10(3): 565–91.
World Bank. 1992. *Strategies for Reducing Poverty in the 1990s.* Washington, D.C.: World Bank.
———. 1997. *Sharing Rising Incomes: Disparities in China.* Washington, D.C.: World Bank.

Statistical Appendix

National Accounts

Balance of Payments

Trade

External and Domestic Debt

Money and Credit

Fiscal Accounts

Agriculture

Industry

Wages

Employment

Prices

Investment

Energy

Transport

Table 1: National Accounts
(billions of yuan, in current prices)

	1985	1986	1987	1988	1989	1990	1991	1992	1993	1994	1995
GDP at market prices	896.4	1,020.2	1,196.3	1,492.8	1,690.9	1,854.8	2,161.8	2,663.8	3,463.4	4,662.2	5826.1
GDP at factor cost	..	..	..	..	..	..	..	..	..	..	..
Agriculture	254.2	276.4	320.4	383.1	422.8	501.7	528.9	580.0	688.2	945.7	1199.3
Industry	386.7	449.3	525.2	658.7	727.8	771.7	910.2	1,170.0	1,642.9	2,237.2	2817.3
Mining and quarrying	10.9	13.9	15.0	18.4	22.2	24.7	28.0	33.6	45.8	62.3	81.5
Manufacturing	317.5	362.4	419.8	532.6	594.5	623.4	734.6	938.3	1,288.9	1,754.7	2193.4
Services	255.6	294.6	350.7	451.0	540.3	581.4	722.7	913.9	1,132.4	1,479.3	1809.4
Imports of goods & non-factor services	125.8	149.8	161.4	205.5	220.0	223.4	289.0	407.1	566.7	961.6	1130.3
Exports of goods & non-factor services	89.1	124.3	162.6	190.4	201.4	274.5	350.7	434.7	498.7	1,025.0	1220.7
Resource balance	-36.7	-25.5	1.2	-15.1	-18.6	51.1	61.7	27.6	-68.0	63.4	90.4
Total expenditures	933.1	1,045.7	1,195.1	1,507.9	1,709.5	1,803.7	2,100.1	2,636.2	3,531.4	4,598.8	5735.7
Total consumption	594.5	661.1	762.9	958.4	1,100.0	1,159.3	1,348.4	1,672.6	2,031.6	2,739.6	3377.0
General government	118.4	136.7	149.0	172.7	203.3	225.2	283.0	349.2	450.0	598.6	712.3
Non-government	476.1	524.4	613.9	785.7	896.7	934.2	1,065.3	1,323.4	1,581.6	2,141.0	2664.7
Statistical discrepancy	17.2	6.9	17.8	22.4	44.3	22.8	33.7	77.4	13.4	18.0	-119.2
Gross domestic investment	338.6	384.6	432.2	549.5	609.5	644.4	751.7	963.6	1,499.8	1,859.2	2358.7
Gross domestic fixed investment	264.1	309.8	374.2	462.4	433.9	473.2	594.0	831.7	1,298.0	1,685.6	2055.4
Nonfinancial public sector	168.1	197.9	229.8	276.3	253.5	291.9	362.8	527.4	765.8	961.6	1089.8
Non-State sector	96.0	112.0	144.4	186.1	180.4	181.3	231.2	304.3	532.2	724.0	965.6
Changes in stocks	74.5	74.8	58.0	87.1	175.6	171.2	157.7	131.9	201.8	173.6	303.3
Gross domestic saving	301.9	359.1	433.4	534.4	591.0	695.5	813.4	991.2	1,431.8	1,922.6	2449.1
Net factor income	2.5	-0.1	-0.8	-0.6	0.9	5.0	4.5	1.4	-7.4	-8.9	-98.3
Net current transfers	0.5	0.9	0.9	1.5	0.9	1.1	2.4	4.4	5.1	7.2	11.7
Gross national saving	304.9	359.9	433.5	535.3	592.7	701.6	820.2	997.0	1,429.5	1,920.9	2362.4
Net indirect taxes	94.2	93.2	93.4	104.9	125.5	133.1	161.3	187.9	295.9	362.0	446.2
Indirect taxes	144.9	151.4	160.4	181.2	222.8	229.1	249.6	264.6	367.0	430.0	517.4
Subsidies	50.7	58.2	67.0	76.3	97.3	96.0	88.3	76.7	71.1	68.0	71.2
Gross national product	898.9	1,020.1	1,195.5	1,492.2	1,691.8	1,859.8	2,166.3	2,665.2	3,456.1	4,653.3	5727.7
Nominal official exchange rate (annual average)	2.94	3.45	3.72	3.72	3.77	4.78	5.32	5.51	5.76	8.62	8.35
GDP at market price (current million US$)	304,912	295,716	268,217	307,167	342,291	354,644	376,617	418,181	431,797	540,944	697614

Source: *China Statistical Yearbook 1996*, pp.42,46.

Table 2: National Accounts
(billions of yuan, in constant 1990 prices)

	1985	1986	1987	1988	1989	1990	1991	1992	1993	1994	1995
GDP at market prices	1,270.2	1,381.9	1,542.2	1,716.5	1,786.9	1,854.8	2,025.4	2,313.0	2,625.3	2,956.1	3266.5
Net indirect taxes	133.5	126.2	120.4	120.6	132.6	133.1	151.1	163.2	224.3	229.5	250.2
GDP at factor cost	..	..	..	..	..	..	..	..	..	..	..
Agriculture	409.1	422.6	442.4	453.5	467.6	501.7	513.7	537.9	563.2	585.7	615.0
Industry	502.2	553.4	629.2	720.4	747.8	771.7	879.0	1,065.4	1,277.4	1,512.4	1725.7
Mining and quarrying	14.1	17.1	18.0	20.1	22.8	24.7	27.1	30.6	35.6	42.1	49.9
Manufacturing	412.3	446.4	503.0	582.5	610.9	623.4	709.4	854.4	1,002.2	1,186.2	1343.5
Services, etc.	358.9	406.0	470.6	542.6	571.5	581.4	632.7	709.8	784.8	858.0	925.8
Imports of goods & non-factor services	252.1	218.5	199.8	238.5	256.4	223.4	258.9	332.2	433.6	473.9	497.4
Exports of goods & non-factor services	156.3	182.5	202.8	228.3	244.5	274.5	316.7	365.3	399.0	510.9	559.9
Resource balance	-95.8	-36.0	3.0	-10.2	-11.9	51.1	57.8	33.1	-34.6	37.0	62.5
Total expenditures	1,366.0	1,417.9	1,539.2	1,726.7	1,798.8	1,803.7	1,967.7	2,280.0	2,659.9	2,919.1	3204.0
Total consumption, etc.	886.2	896.9	982.0	1,094.9	1,154.7	1,159.3	1,263.4	1,489.6	1,673.8	1,829.0	1905.5
General government	181.7	187.5	196.4	202.6	222.9	225.1	271.6	326.2	373.3	402.4	388.7
Non-government	704.5	709.5	785.6	892.3	931.8	934.2	991.7	1,163.4	1,300.5	1,426.6	1516.8
Gross domestic investment /a	479.8	521.0	557.2	631.8	644.1	644.4	704.3	790.4	986.1	1,090.1	1298.5
Gross domestic fixed investment /a	374.2	419.6	482.4	531.7	458.5	473.2	556.5	675.8	833.1	980.0	1128.4
Nonfinancial public sector /a	237.8	267.7	295.8	317.1	267.5	291.9	339.8	458.1	580.2	609.3	611.5
Non-State sector /a	136.4	151.9	186.6	214.5	191.0	181.3	216.7	217.7	252.9	370.7	516.9
Changes in stocks	105.6	101.3	74.8	100.2	185.6	171.2	147.8	114.5	153.0	110.1	170.0
Net factor income	2.0	-1.6	-3.2	-3.6	-2.0	5.0	3.7	2.1	-6.8	-7.7	-52.7
Net current transfers	0.7	1.2	1.2	1.8	0.9	1.1	2.2	3.8	3.8	4.6	6.5
Gross national product	1,272.2	1,380.3	1,539.0	1,712.9	1,784.9	1,859.8	2,029.1	2,315.2	2,618.5	2,948.4	3213.8
Gross domestic saving	406.2	483.7	558.6	614.3	622.5	695.5	759.6	812.9	934.1	1,121.3	1338.2
Gross national saving	409.0	483.3	556.6	612.5	621.4	701.6	765.4	818.9	931.1	1,118.2	1292.0
Capacity to import	178.6	181.3	201.2	221.0	234.8	274.5	314.2	354.7	381.5	505.2	537.2
Terms of trade adjustment	22.3	-1.3	-1.6	-7.3	-9.7	0.0	-2.5	-10.6	-17.4	-5.8	-22.7
Gross domestic income	1,292.4	1,380.7	1,540.7	1,709.2	1,777.2	1,854.8	2,022.9	2,302.5	2,607.9	2,950.3	3243.7
Gross national income	1,294.4	1,379.0	1,537.4	1,705.6	1,775.1	1,859.8	2,026.6	2,304.6	2,601.0	2,942.6	3191.0

/a In the absence of an official fixed investment deflator, the real investment numbers are derived by employing the GDP deflator until 1991.
Thereafter the fixed investment deflator is employed. (*China Statistical Yearbook* , p. 272)

Source: China Statistical Yearbook 1996, pp.42,46.

Table 3: National Accounts
(implicit price deflators, 1990=100)

	1985	1986	1987	1988	1989	1990	1991	1992	1993	1994	1995
GDP at market prices	70.6	73.8	77.6	87.0	94.6	100.0	106.7	115.2	131.9	157.7	178.4
Net indirect taxes	70.6	73.8	77.6	87.0	94.6	100.0	106.7	115.2	131.9	157.7	178.4
GDP at factor cost	..	..	..	..	..	..	..	..	..	..	..
Agriculture	62.1	65.4	72.4	84.5	90.4	100.0	102.9	107.8	122.2	161.5	195.0
Industry	77.0	81.2	83.5	91.4	97.3	100.0	103.6	109.8	128.6	147.9	163.3
Mining and quarrying	77.0	81.2	83.5	91.4	97.3	100.0	103.6	109.8	128.6	147.9	163.3
Manufacturing	77.0	81.2	83.5	91.4	97.3	100.0	103.6	109.8	128.6	147.9	163.3
Services, etc.	71.2	72.6	74.5	83.1	94.5	100.0	114.2	128.8	144.3	172.4	195.4
Imports of goods & non-factor services	49.9	68.6	80.8	86.2	85.8	100.0	111.6	122.5	130.7	202.9	227.2
Exports of goods & non-factor services	57.0	68.1	80.2	83.4	82.4	100.0	110.7	119.0	125.0	200.6	218.0
Terms of trade (Px/Pm)	114.2	99.3	99.2	96.8	96.0	100.0	99.2	97.1	95.6	98.9	95.9
Total expenditures	68.3	73.8	77.6	87.3	95.0	100.0	106.7	115.6	132.8	157.5	179.0
Total consumption, etc.	67.1	73.7	77.7	87.5	95.3	100.0	106.7	112.3	121.4	149.8	177.2
General government	65.2	72.9	75.9	85.3	91.2	100.0	104.2	107.1	120.6	148.8	183.2
Non-government	67.6	73.9	78.1	88.1	96.2	100.0	107.4	113.8	121.6	150.1	175.7
Gross domestic investment	70.6	73.8	77.6	87.0	94.6	100.0	106.7	121.9	152.1	170.6	181.7
Gross domestic fixed investment /a	70.6	73.8	77.6	87.0	94.6	100.0	106.7	123.1	155.8	172.0	182.1
Nonfinancial public sector	70.7	73.9	77.7	87.1	94.8	100.0	106.8	115.1	132.0	157.8	178.2
Non-State sector	70.4	73.7	77.4	86.8	94.4	100.0	106.7	139.8	210.4	195.3	186.8
Changes in stocks	70.6	73.8	77.6	87.0	94.6	100.0	106.7	115.2	131.9	157.7	178.4
Net factor income	123.7	4.9	24.9	16.8	-42.7	100.0	122.3	64.3	108.3	116.3	186.5
Net current transfers	68.3	73.8	77.6	87.3	95.0	100.0	106.7	115.6	132.8	157.5	179.0
Gross national product	70.7	73.9	77.7	87.1	94.8	100.0	106.8	115.1	132.0	157.8	178.2
Gross domestic saving	74.3	74.2	77.6	87.0	94.9	100.0	107.1	121.9	153.3	171.5	183.0
Gross national saving	74.6	74.5	77.9	87.4	95.4	100.0	107.2	121.8	153.5	171.8	182.8

/a In the absence of an official fixed investment deflator, the real investment numbers are derived by employing the GDP deflator until 1991. Thereafter the fixed investment deflator is employed. (*China Statistical Yearbook* , p. 272)

Source: Table 1 divided by Table 2.

Table 4: National Accounts
(percentage growth rates in constant 1990 prices)

	1985	1986	1987	1988	1989	1990	1991	1992	1993	1994	1995
GDP at market prices	13.5	8.8	11.6	11.3	4.1	3.8	9.2	14.2	13.5	12.6	10.5
Net indirect taxes	..	..	..	..	..	..	..	..	..	..	..
GDP at factor cost	..	..	..	..	..	..	..	..	..	..	..
Agriculture	1.8	3.3	4.7	2.5	3.1	7.3	2.4	4.7	4.7	4.0	5.0
Industry	18.6	10.2	13.7	14.5	3.8	3.2	13.9	21.2	19.9	18.4	14.1
Mining and quarrying	13.9	21.2	5.1	11.7	13.4	8.4	9.7	13.2	16.3	..	..
Manufacturing	18.1	8.3	12.7	15.8	4.9	2.0	13.8	20.4	17.3	18.4	13.3
Services, etc.	22.2	13.1	15.9	15.3	5.3	1.7	8.8	12.2	10.6	9.3	7.9
Imports of goods & non-factor services	53.7	-13.3	-8.6	19.4	7.5	-12.9	15.9	28.3	30.5	9.3	5.0
Exports of goods & non-factor services	7.6	16.8	11.1	12.6	7.1	12.3	15.4	15.3	9.2	28.1	9.6
Resource balance	..	..	..	..	..	..	..	..	..	..	..
Total expenditures	20.0	3.8	8.6	12.2	4.2	0.3	9.1	15.9	16.7	9.7	9.8
Total consumption, etc.	17.7	1.2	9.5	11.5	5.5	0.4	9.0	17.9	12.4	9.3	4.2
General government	11.2	3.2	4.8	3.1	10.0	1.0	20.7	20.1	14.4	7.8	-3.4
Non-government	19.6	0.7	10.7	13.6	4.4	0.3	6.2	17.3	11.8	9.7	6.3
Gross domestic investment	24.5	8.6	7.0	13.4	1.9	0.0	9.3	12.2	24.8	10.5	19.1
Gross domestic fixed investment	12.8	12.1	15.0	10.2	-13.8	3.2	17.6	21.4	23.3	17.6	15.1
Nonfinancial public sector	28.6	12.6	10.5	7.2	-15.7	9.1	16.4	34.8	26.7	5.0	0.4
Non-State sector	-7.1	11.4	22.8	15.0	-11.0	-5.1	19.5	0.5	16.2	46.6	39.4
Changes in stocks	97.2	-4.0	-26.2	33.9	85.3	-7.7	-13.7	-22.5	33.6	-28.0	54.5
Net factor income	..	..	..	..	..	..	..	..	..	..	..
Net current transfers	..	..	..	..	..	..	..	..	..	..	..
Gross national product	13.2	8.5	11.5	11.3	4.2	4.2	9.1	14.1	13.1	12.6	9.0
Gross domestic saving	5.4	19.1	15.5	10.0	1.3	11.7	9.2	7.0	14.9	20.0	19.3
Gross national saving	4.5	18.2	15.2	10.0	1.5	12.9	9.1	7.0	13.7	20.1	15.5
Capacity to import	..	..	..	..	..	..	..	..	..	..	..
Terms of trade adjustment	..	..	..	..	..	..	..	..	..	..	..
Gross domestic income	13.5	6.8	11.6	10.9	4.0	4.4	9.1	13.8	13.3	13.1	9.9
Gross national income	13.3	6.5	11.5	10.9	4.1	4.8	9.0	13.7	12.9	13.1	8.4

Source: Table 2

Table 5: Balance of Payments
(billions of US dollars)

	1985	1986	1987	1988	1989	1990	1991	1992	1993	1994	1995	1996
Exports of goods and non-factor services	28.2	29.6	39.1	45.9	47.8	57.3	65.8	78.8	86.5	118.8	147.2	153.7
Merchandise (fob)	25.1	25.8	34.7	41.1	43.2	51.5	58.9	69.6	75.7	102.6	128.1	128.5
Non-factor services	3.1	3.8	4.4	4.8	4.6	5.8	6.9	9.2	10.9	16.3	19.1	25.2
Imports of goods and non-factor services	41.1	37.5	38.9	50.0	52.7	46.6	54.3	73.8	98.3	111.5	135.3	141.3
Merchandise (fob)	38.2	34.9	36.4	46.4	48.8	42.4	50.2	64.4	86.3	95.3	110.1	114.6
Non-factor services	2.9	2.6	2.5	3.6	3.9	4.3	4.1	9.4	12.0	16.2	25.2	26.7
Resource balance	-13.0	-7.9	0.2	-4.1	-5.0	10.7	11.5	5.0	-11.8	7.3	12.0	12.4
Net factor income	0.9	0.2	-0.2	-0.1	0.3	1.0	0.9	0.3	-1.3	-1.0	-11.8	-10.4
Factor receipts	1.5	1.1	1.0	1.5	1.9	3.1	3.8	5.7	4.4	5.9	5.2	7.4
Factor payments	0.5	0.9	1.2	1.6	1.7	2.1	2.9	5.4	5.7	6.9	17.0	17.8
Total interest due	1.2	1.1	1.8	2.2	3.2	3.1	3.7	3.4	3.4	4.8	6.0	..
Other factor payments & disc.	-0.6	-0.1	-0.6	-0.6	-1.5	-1.1	-0.8	2.0	2.3	2.1	11.0	17.8
Net current transfers	0.2	0.3	0.2	0.4	0.2	0.2	0.4	0.8	0.9	0.8	0.8	1.6
Current receipts	0.2	0.3	0.3	0.4	0.2	0.2	0.5	0.8	0.9	1.1	1.2	..
Workers remittances	0.2	0.2	0.2	0.1	0.1	0.1	0.2	0.2	0.1	0.4	0.4	..
Other current transfers	0.0	0.1	0.1	0.3	0.2	0.1	0.3	0.6	0.8	0.7	0.8	..
Current payments	0.0	0.0	0.0	0.0	0.0	0.0	0.0	0.0	0.0	0.3	0.4	..
Current account balance before official grants	-11.9	-7.5	0.3	-3.8	-4.5	11.9	12.9	6.1	-12.2	7.2	1.0	3.6
Current account balance as a share of GDP (percent)	-3.9	-2.5	0.1	-1.2	-1.3	3.4	3.4	1.4	-2.8	1.3	0.1	0.5
Official capital grants	0.1	0.1	0.0	0.0	0.1	0.1	0.4	0.4	0.3	0.5	0.6	0.3
Current account balance after official grants	-11.8	-7.3	0.3	-3.8	-4.3	12.0	13.3	6.4	-11.9	7.7	1.6	3.9
Long term capital inflows	5.0	6.3	7.8	9.1	8.7	9.0	8.0	18.4	35.9	41.6	46.7	39.9
Direct investment	1.0	1.4	1.7	2.3	2.6	2.7	3.5	7.2	23.1	31.8	33.8	38.8
Net long term borrowing	4.0	4.9	6.1	6.8	6.1	6.3	4.5	11.3	12.8	9.8	12.9	1.1
Disbursements	5.3	6.7	8.0	9.1	8.4	9.7	8.7	16.5	19.6	16.2	22.0	
Repayments due	1.3	1.9	2.0	2.3	2.4	3.3	4.1	5.2	6.7	6.3	9.1	
Total other items (net)	4.4	-1.0	-3.3	-2.9	-4.9	-9.0	-6.7	-26.9	-22.3	-18.8	-25.9	-12.2
Net short-term capital	2.3	-2.3	0.2	0.1	-1.5	-3.2	0.4	-0.9	-3.9	-3.1	0.4	0.1
Capital flows not elsewhere included	0.0	0.0	0.0	0.0	0.0	-2.6	-0.3	-17.8	-8.5	-5.8	-8.5	2.4
Errors and omissions	2.1	1.3	-3.5	-3.0	-3.3	-3.2	-6.8	-8.2	-9.8	-9.8	-17.8	-14.7
Changes in net reserves	2.4	2.0	-4.8	-2.4	0.5	-12.0	-14.5	2.1	-1.8	-30.5	-22.5	-31.6
Net credit from IMF	0.0	0.7	0.1	0.0	0.0	-0.4	-0.4	0.0	0.0	0.0	0.0	0.0
Reserve changes not elsewhere included	2.4	1.3	-4.9	-2.3	0.5	-11.6	-14.1	2.1	-1.8	-30.5	-22.5	-31.6
Escrow account	0.0	0.0	0.0	0.0	0.0	0.0	0.0	0.0	0.0	0.0	0.0	0.0
Gross reserves (excluding gold) /a	12.7	11.5	16.3	18.5	18.0	29.6	43.7	20.6	22.4	52.9	75.4	107.7
Gross reserves (including gold) /b	16.9	16.4	22.5	23.8	23.1	34.5	48.3	25.0	27.0	57.8	80.3	..
Exchange rates:												
Nominal official exchange rate (average)	2.9	3.5	3.7	3.7	3.8	4.8	5.3	5.5	5.8	8.6	8.4	8.3
Nominal official exchange rate (end-of-year)	3.2	3.7	3.7	3.7	4.7	5.2	5.4	5.8	5.8	8.4	8.3	8.3
Manufactures Unit Value Index (% change)	0.8	17.9	9.8	7.3	-0.7	7.8	2.2	4.3	-0.3	3.7	8.3	-2.5
Real effective exchange rate index	63.6	46.3	40.2	43.7	50.7	37.3	32.4	31.3	30.7	33.5	35.3	37.0

/a Since August 1992 the authorities have defined gross international reserves as the sum of only state foreign exchange reserves (not total reserves), gold, reserve position in the Fund and SDR holdings.

/b Gold valued at London prices (Source: IFS)

Source: World Bank, IMF: *International Financial Statistics*.

Table 6: Services
(millions of US dollars)

	1985	1986	1987	1988	1989	1990	1991	1992	1993	1994	1995	1996
A. Shipment of freight												
Credit	671	705	904	1,308	1,061	1,937	1,179	1,294	1,391	2,065	2,478	..
Debit	1,224	850	1,186	1,387	2,382	2,139	2,193	3,876	5,134	6,926	8,727	..
B. Insurance												
Credit	196	229	252	345	332	227	342	486	452	1,700	1,852	..
Debit	69	82	142	214	187	84	214	274	362	1,880	4,273	..
C. Other transportation												
Credit	271	304	152	169	153	480	494	0	0	0	0	..
Debit	0	0	0	0	0	0	0	0	0	0	0	..
D. Port expenses												
Credit	360	306	289	304	300	289	338	368	245	760	874	910
Debit	300	670	456	889	370	1,106	314	449	345	695	799	910
E. Travel receipts												
Credit	979	1,227	1,693	2,078	1,707	1,738	2,346	3,530	4,683	7,323	8,730	10,200
Debit	314	308	387	633	429	470	511	2,512	2,797	3,036	3,688	4,430
F. Profits												
Credit	6	0	10	0	6	0	0	0	0	0	1	0
Debit	14	15	2	8	7	46	10	21	231	400	9,965	10,960
G. Interest												
Credit	484	216	177	427	247	667	747	610	535	928	1,104	1,870
Debit	68	298	457	644	394	480	870	1,890	2,079	2,506	3,150	2,940
H. Bank interest and charges												
Credit	897	685	789	1,042	1,641	2,350	2,972	4,985	3,855	4,809	4,086	5,560
Debit	464	611	732	978	1,264	1,536	1,999	3,436	3,364	3,868	3,850	3,900
I. Posts												
Credit	13	15	12	24	118	159	221	349	471	706	756	..
Debit	7	15	14	11	16	13	15	72	85	146	217	..
J. Interofficial												
Credit	130	215	204	137	151	107	115	141	201	266	700	..
Debit	263	251	150	277	337	239	184	227	473	518	588	..
K. Labor income												
Credit	91	199	51	35	53	52	74	60	47	117	..	..
Debit	0	0	0	0	0	0	0	20	22	98	..	..
L. Other services												
Credit	435	940	880	458	727	866	1,870	2,604	3,409	3,429	3,740	..
Debit	347	100	150	193	189	291	689	2,004	2,818	3,000	6,931	..
M. Total services												
Net	1,069	1,427	1,737	1,094	923	2,558	3,697	63	-2,421	-969	-17,868	-11,880
Credit	4,533	4,927	5,413	6,327	6,497	8,872	10,697	14,844	15,289	22,104	24,321	32,630
Debit	3,464	3,500	3,676	5,233	5,574	6,314	7,000	14,781	17,710	23,073	42,189	44,510
N. factor Services (F+G+H+K)												
Net	932	176	-164	-126	282	1,007	914	288	-1,259	-1,018	-11,774	-10,370
Receipts (credit)	1,478	1,100	1,027	1,504	1,947	3,069	3,793	5,655	4,437	5,854	5,191	7,430
Payments (debit)	546	924	1,191	1,630	1,665	2,062	2,879	5,367	5,696	6,872	16,965	17,800
O. Non-factor services (A+B+C+D+E+I+L)												
Net	137	1,251	1,901	1,220	641	1,551	2,783	-225	-1,162	49	-6,094	-1,510
Receipts (credit)	3,055	3,827	4,386	4,823	4,550	5,803	6,904	9,189	10,852	16,250	19,130	25,200
Payments (debit)	2,918	2,576	2,485	3,603	3,909	4,252	4,121	9,414	12,014	16,201	25,224	26,710

Source: World Bank; IMF: *International Financial Statistics*.

Table 7: Transfers
(millions of US dollars)

	1985	1986	1987	1988	1989	1990	1991	1992	1993	1994	1995	1996
Private unrequited transfers												
Net	171	255	249	416	238	222	444	804	883	836	810	1,580
Credit	180	266	260	428	247	233	484	821	901	1,095	1,170	..
Debit	9	11	11	12	9	11	40	17	18	259	360	..
Nonresidential remittances												
Net	177	205	163	125	73	119	189	213	93	376	350	..
Credit	180	208	166	129	76	124	207	228	108	395	350	..
Debit	3	3	3	4	3	5	18	15	15	19	0	..
Migrants' transfers												
Net	-6	50	86	291	165	103	255	591	790	460	460	..
Credit	0	58	94	299	171	109	277	593	793	700	820	..
Debit	6	8	8	8	6	6	22	2	3	240	360	..
Public unrequited transfers												
Net	73	124	-25	3	143	52	387	351	289	501	625	300
Credit	260	250	129	140	230	143	406	385	389	674	657	..
Debit	187	126	154	137	87	91	19	34	100	173	32	..
International organizations												
Net	21	96	24	42	82	69	183	139	127	121	154	..
Credit	63	140	58	61	120	84	192	173	160	144	186	..
Debit	42	44	34	19	38	15	9	34	33	23	32	..
Grants and aid												
Net	52	28	-49	-39	61	-17	204	212	162	380	471	..
Credit	197	110	71	79	110	59	214	212	229	530	471	..
Debit	145	82	120	118	49	76	10	0	67	150	0	..
Total transfers												
Net	243	379	224	419	381	274	831	1,155	1,172	1,337	1,435	1,880
Credit	439	516	389	568	477	376	890	1,206	1,290	1,769	1,827	..
Debit	196	137	165	149	96	102	59	51	118	432	392	..

Source: World Bank, IMF: *International Financial Statistics.*

Table 8: International Reserves (millions of US dollars)

	1985	1986	1987	1988	1989	1990	1991	1992	1993	1994	1995	1996
Total reserves (minus gold)	12,728	11,453	16,305	18,541	17,960	29,586	43,674	20,620	22,387	52,914	75,377	107,652
SDR's	483	569	640	586	540	562	577	419	484	539	582	614
Reserve position with Fund	332	370	429	407	398	430	433	758	704	755	1,216	1,396
Foreign exchange reserves	11,913	10,514	15,236	17,548	17,022	28,594	42,664	19,443	21,199	51,620	73,579	10,500
Gold												
Gold (million fine troy ounces)	13	13	13	13	13	13	13	13	13	13	13	..
Gold (National valuation) /a	486	541	629	594	587	623	634	610	612	646	660	637
London gold price(US$ per oz) /b	327	391	484	410	401	384	362	343	360	384	384	388
Gold at London price (US$ million)	4,153	4,964	6,148	5,210	5,093	4,871	4,600	4,361	4,569	4,880	4,879	..
Total reserves including gold (National valulation)	13,214	11,994	16,934	19,135	18,547	30,209	44,308	21,230	22,999	53,560	76,037	108,289
Total reserves including Gold (London price)	16,881	16,417	22,453	23,751	23,053	34,457	48,274	24,981	26,956	57,794	80,256	..

/a From August 1992 onwards the authorities have defined gross international reserves as the sum of only state foreign exchange reserves (not total reserves), gold, reserve position with the Fund and SDR holdings.

/b Gold valued at SDR 35 per fine ounce.

Source: World Bank, IMF: *International Financial Statistics.*

Table 9: Commodity Composition of Merchandise Exports (millions of US dollars, customs basis)

	1985	1986	1987	1988	1989	1990	1991	1992	1993	1994	1995
PRIMARY GOODS	13,828	11,272	13,231	14,406	15,078	15,886	16,145	17,004	16,666	19,708	21,487
FOOD	3,803	4,448	4,781	5,890	6,145	6,609	7,226	8,309	8,399	10,015	9,954
Live animals chiefly for food	304	338	348	386	395	430	439	479	453	468	503
Meat and meat products	448	483	520	585	657	791	906	770	950	909	1,371
Fishes, shell-fish, molluscs etc.	283	491	721	969	1,039	1,370	1,181	1,366	1,254	2,320	2,853
Grain and grain products	1,065	898	579	681	719	614	1,169	1,692	1,660	1,687	285
Vegetables and fruits	825	1,092	1,290	1,617	1,623	1,742	1,946	2,023	2,163	2,889	3,342
Coffee, tea, cocoa etc.	435	466	488	524	568	534	491	499	510	484	516
NON-FOOD	2,653	2,908	3,650	4,257	4,212	3,537	3,486	3,143	3,052	4,127	4,375
Oil seeds and oil-containing fruits	487	580	674	684	645	619	741	867	793	666	522
Textile fibers etc.	1,145	1,160	1,508	1,672	1,546	1,095	1,125	4,224	4,179	1,093	753
Animal and vegetable raw materials	398	486	645	724	845	809	705	606	617	1,136	1,351
MINERAL FUELS	7,132	3,683	4,544	3,950	4,321	5,237	4,754	4,693	4,109	4,069	5,335
Coal, coke and briquettes	349	455	536	594	680	755	829	..	..	1,054	1,695
Petroleum, petroleum products etc.	6,777	3,224	4,003	3,350	3,633	4,460	3,975	..	..	2,789	3,243
OTHER	240	233	256	309	400	503	679	859	1,106	1,497	1,823
MANUFACTURED GOODS	13,522	19,670	26,206	33,110	37,460	46,205	55,698	67,936	75,078	101,298	127,283
CHEMICALS AND RELATED PRODUCTS	1,358	1,733	2,235	2,897	3,201	3,730	3,818	4,348	4,623	6,236	9,094
Organic	309	411	500	575	690	838	911	1,403	1,541	1,602	2,285
Inorganic	287	379	553	762	791	842	913	1,050	1,145	1,350	2,225
LIGHT INDUSTRY	4,493	5,886	8,570	10,489	10,897	12,576	14,456	16,135	16,392	23,218	32,243
Yarn, fabrics, manufactured goods etc.	3,243	4,220	5,790	6,456	6,994	6,999	7,734	..	..	11,818	13,919
Non-metallic minerals	227	317	439	579	792	1,316	1,668	..	..	2,521	3,425
Metal products	426	553	797	1,006	1,210	1,283	1,669	..	..	1,654	5,225
MACHINERY AND TRANSPORT EQUIPMENT	772	1,094	1,741	2,769	3,874	5,588	7,149	13,219	15,282	21,895	31,391
OTHER	3,486	4,948	6,273	8,268	10,755	12,686	16,620	34,234	38,781	49,937	54,548
Clothing and garments	2,050	2,913	3,749	4,872	6,130	6,848	8,998	16,883	18,325	23,732	24,049
PRODUCTS NOT CLASSIFIED ELSEWHERE	3,413	6,009	7,387	8,687	8,733	11,625	13,655	..	..	12	7
TOTAL	27,350	30,942	39,437	47,516	52,538	62,091	71,843	84,940	91,744	121,006	148,770

Note : Data from 1985-91 are based on Standard Industrial Trade Classification (SITC); 1992-95 categories are based on the Harmonized System (HS).
From 1992, Customs Statistics use new commodity catagories; products not otherwise classified have been included in different categories of commodities.
Source: *China Statistical Yearbook* 1996, p.581.

Table 10: Imports (millions of US dollars, customs basis)

	1985	1986	1987	1988	1989	1990	1991	1992	1993	1994	1995
FOOD	1,881	2,002	3,055	4,191	5,269	4,474	3,718	3,907	2,953	5,014	9,126
Food	1,553	1,625	2,443	3,476	4,192	3,335	2,799	3,143	2,206	3,137	6,131
Beverages	206	172	263	346	202	157	200	239	245	68	394
Animal fat	122	205	349	369	875	982	719	525	502	1,809	2,601
PETROLEUM (Mineral fuels)	172	504	539	787	1,650	1,272	2,113	3,570	5,819	4,035	5,127
INTERMEDIATE	19,175	17,552	16,769	23,391	23,415	18,325	23,189	34,237	40,042	43,955	52,066
Chemicals and related products	4,469	3,771	5,008	9,139	7,556	6,648	9,277	11,157	9,704	12,130	17,300
Crude materials (non-food)	3,236	3,143	3,321	5,090	4,835	4,107	5,003	5,775	5,438	7,437	10,158
Leather and Cork	770	851	728	842	747	938	1,267	1,626	1,859	2,943	3,020
Leather	..	..	184	224	280	374	642	206	263	1,902	1,993
Cork	..	..	544	618	467	564	625	1,420	1,596	1,041	1,027
Textile yarn (yarn, fabrics etc.)	1,607	1,632	1,848	2,388	2,845	2,748	3,689	3,690	3,145	9,347	10,914
Non metallic minerals	325	363	342	430	520	453	443	4,519	3,776	1,001	1,113
Iron and steel	7,120	6,741	4,787	4,624	5,797	2,852	2,694	5,051	13,896	9,438	6,878
Non-ferrous metals	1,648	1,051	735	878	1,114	579	816	2,420	2,224	1,659	2,683
CONSUMER GOODS	2,330	2,431	1,743	1,757	1,866	2,051	2,506	7,949	8,666	6,987	8,277
Paper (paper and related products)	428	554	727	610	634	745	969	1,771	1,741	1,923	2,157
Rubber	1,902	1,877	45	51	50	50	76	555	598	186	224
Furniture	..	..	42	61	68	72	49	178	218	111	90
Travel goods	..	..	3	8	6	6	7	302	327	50	42
Clothing	..	..	17	28	38	48	61	437	543	622	969
Footwear	..	..	1	2	3	9	11	506	513	325	341
Photo supplies	..	..	432	365	398	361	441	2,024	2,320	1,656	1,861
Miscellaneous	..	..	476	632	669	759	892	2,177	2,407	2,114	2,592
MANUFACTURED	18,694	20,415	21,110	25,150	26,940	27,223	32,264	30,922	46,479	55,624	57,481
Total	42,252	42,904	43,216	55,275	59,140	53,345	63,791	80,585	103,959	115,614	132,078

Note : Data from 1985-91 are based on Standard Industrial Trade Classification (SITC); 1992-95 categories are based on the Harmonized System (HS).

Source : *China Statistical Yearbook* 1996, p. 581.

Table 11: External Debt: Disbursements and Repayments
(millions of US dollars)

	1985	1986	1987	1988	1989	1990	1991	1992	1993	1994	1995
DISBURSEMENTS											
Public & publicly guaranteed long-term debt	5,280	6,732	8,044	9,065	8,442	9,665	8,659	16,308	19,229	16,151	21,411
Official creditors	1,166	1,444	1,123	1,847	2,761	2,578	2,649	3,103	5,501	4,200	8,373
Multilateral	599	620	717	1,124	1,169	1,158	1,455	1,523	2,252	2,558	2,838
of which IDA	212	282	399	557	507	507	612	778	869	680	812
of which IBRD	354	324	303	553	604	591	668	552	977	1,380	1,457
Bilateral	567	824	405	724	1,592	1,420	1,194	1,580	3,248	1,642	5,535
Private creditors	4,114	5,288	6,921	7,218	5,681	7,087	6,010	13,204	13,729	11,951	13,038
Bonds	971	1,333	1,064	782	450	277	260	894	2,737	3,337	1,224
Commercial banks	700	1,782	4,605	4,470	2,016	3,247	2,623	5,062	5,624	2,380	4,977
Other private	2,443	2,172	1,252	1,967	3,206	3,564	3,127	7,248	5,368	6,234	6,837
Private non-guaranteed long-term	0	0	0	0	0	0	0	198	332	0	544
Total long-term disbursements	5,280	6,732	8,044	9,065	8,442	9,665	8,659	16,505	19,561	16,151	21,955
IMF purchases	0	701	0	0	0	0	0	0	0	0	0
Net short-term capital	..	..	..	..	..	..	..	..	..	..	..
Total disbursements	5,280	6,732	8,044	9,065	8,442	9,665	8,659	16,505	19,561	16,151	21,955
REPAYMENT DUE											
Public & publicly guaranteed long-term debt	1,297	1,874	1,956	2,285	2,365	3,319	4,123	5,213	6,729	6,343	9,070
Official creditors	49	279	496	492	485	851	605	760	886	1,083	1,171
Multilateral	0	2	99	41	63	220	141	215	272	359	420
of which IDA	0	0	0	0	0	0	1	2	4	9	14
of which IBRD	0	0	97	39	62	216	130	196	245	315	350
Bilateral	49	277	397	451	421	632	464	545	614	725	761
Private creditors	1,248	1,595	1,460	1,793	1,880	2,468	3,517	4,453	5,843	5,260	7,899
Bonds	0	0	0	11	33	325	236	1,095	831	461	1,451
Commercial banks	77	331	466	754	867	808	2,010	2,046	2,895	1,803	2,645
Other private	1,171	1,264	993	1,028	980	1,335	1,272	1,311	2,117	2,996	3,803
Private non-guranteed long-term debt	0	0	0	0	0	0	0	0	0	0	0
Total long-term repayments due	1,297	1,874	1,956	2,285	2,365	3,319	4,123	5,213	6,729	6,343	9,070
IMF repurchases	0	36	81	83	79	490	451	0	0	0	0
Total long-term repayment & IMF repurchase	1,297	1,910	2,037	2,368	2,444	3,809	4,574	5,213	6,729	6,343	9,070
NET FLOWS											
Official creditors	1,117	1,165	626	1,355	2,277	1,727	2,044	2,343	4,615	3,117	7,202
of which IDA	212	282	399	557	507	507	611	777	865	671	798
of which IBRD	354	324	206	514	542	376	538	357	732	1,065	1,107
COMMITMENTS											
IBRD commitments	660	672	692	868	1,221	75	1,312	1,253	1,445	2,930	2,495
of which fast disbursing	0	0	0	200	0	0	0	0	0	0	0
IDA commitments	433	448	613	594	539	878	1,310	612	870	1,090	355
of which fast disbursing	0	0	0	97	0	0	0	0	0	0	0

Source : World Bank: *World Debt Tables*

Table 12: External Debt: Interest and Debt Outstanding
(millions of US dollars)

	1985	1986	1987	1988	1989	1990	1991	1992	1993	1994	1995
INTEREST DUE											
Public & publicly guaranteed long-term debt	586	645	1,125	1,611	2,511	2,534	2,953	2,708	2,618	3,818	4,623
Official creditors	172	262	402	434	457	531	635	678	827	1,131	1,288
Multilateral	30	76	126	143	179	226	263	319	376	480	619
of which IDA	4	8	12	15	14	19	23	29	34	41	49
of which IBRD	26	66	111	126	161	200	227	264	299	364	460
Bilateral	142	187	276	290	278	305	372	358	450	651	669
Private creditors	414	382	723	1,178	2,054	2,003	2,319	2,030	1,792	2,687	3,336
Bonds	20	91	213	289	347	367	356	337	286	363	594
Commercial banks	50	67	137	457	1,062	959	1,071	776	738	1,034	1,333
Other private	343	225	373	432	646	677	891	918	767	1,289	1,408
Private non-guaranteed long-term debt	0	0	0	0	0	0	0	0	12	26	33
Interest arrears	0	0	0	0	0	0	0	0	0	0	0
Reduction in arrears (-)	0	0	0	0	0	0	0	0	0	0	0
Total long-term interest due	586	645	1,125	1,611	2,511	2,534	2,953	2,708	2,630	3,844	4,657
IMF service charges	2	2	50	51	67	65	24	0	0	0	0
Interest on short-term debt	594	417	640	534	628	547	707	697	809	948	1,340
Total interest due	1,181	1,064	1,815	2,197	3,206	3,146	3,684	3,405	3,439	4,792	5,997
DEBT OUTSTANDING AND DISBURSED (DOD)											
Public & publicly guaranteed long-term debt	9,937	16,571	25,963	32,620	37,118	45,515	49,479	58,463	70,076	82,391	94,674
Official creditors	4,724	7,028	9,496	10,536	12,039	14,514	17,073	19,105	24,339	28,973	36,282
Multilateral	983	1,810	2,852	3,753	4,783	6,111	7,576	8,614	10,690	13,588	16,302
of which IDA	431	774	1,330	1,819	2,296	3,016	3,672	4,286	5,160	6,097	7,038
of which IBRD	498	965	1,427	1,831	2,330	2,865	3,494	3,752	4,549	5,933	7,209
Bilateral	3,741	5,218	6,644	6,783	7,257	8,403	9,497	10,491	13,650	15,385	19,980
Private creditors	5,213	9,544	16,467	22,085	25,079	31,001	32,406	39,358	45,737	53,418	58,393
Bonds	1,234	2,811	4,498	5,182	5,228	5,425	5,660	5,449	7,715	11,087	10,684
Commercial banks	776	1,780	6,087	10,393	11,432	14,520	14,963	17,913	20,678	21,475	23,869
Other private	3,203	4,953	5,882	6,509	8,419	11,055	11,783	15,995	17,344	20,856	23,840
Private non-guaranteed long-term	0	0	0	0	0	0	0	200	556	583	1,090
Total long-term DOD	9,937	16,571	25,963	32,620	37,118	45,515	49,479	58,663	70,632	82,974	95,764
Use of IMF credit	340	1,072	1,155	1,013	908	469	0	0	0	0	0
Short-term debt	6,419	6,076	8,221	8,806	6,907	9,317	10,780	13,765	15,296	17,483	22,325
Total external debt	16,696	23,719	35,340	42,439	44,932	55,301	60,259	72,428	85,928	100,457	118,089
MEMORANDUM ITEMS											
% Debt on concessional terms	25	22	20	19	21	21	21	20	19	19	18
% Debt at variable interest rates	24	25	34	39	38	36	33	28	29	28	30
% Bilateral debt on concessional terms	12	12	11	10	12	13	11	10	12	12	11
% Multilateral debt on concessional terms	3	3	4	4	5	6	6	6	8	8	8
Preferred creditor debt service	2	5	11	8	8	17	12	7	7	8	8

Source : World Bank, *World Debt Tables*.

Table 13: Domestic Debt
(billions of yuan)

Year issued	Type	Issued to	Amount issued	1987	1988	1989	1990	1991	1992	1993	1994	1995	1996	1997	1998	1999	2000
1982	Treasury bonds	Enterprises	2.40	0.48	0.48	0.48	0.48										
	Treasury bonds	Households	2.00	0.39	0.39	0.39	0.39										
1983	Treasury bonds	Enterprises	2.10		0.42	0.42	0.42	0.42	0.42								
	Treasury bonds	Households	2.10		0.42	0.42	0.42	0.42	0.42								
1984	Treasury bonds	Enterprises	2.00			0.40	0.40	0.40	0.40	0.40							
	Treasury bonds	Households	2.20			0.44	0.44	0.44	0.44	0.44							
1985	Treasury bonds	Enterprises	2.20				2.18										
	Treasury bonds	Households	3.80				3.75										
1986	Treasury bonds	Enterprises	2.30					2.29									
	Treasury bonds	Households	4.00					3.96									
1987	Treasury bonds	Enterprises	2.30						2.26								
	Treasury bonds	Households	4.00						4.00								
	Key construction bonds	Households	0.50				0.50										
	Key construction bonds	Enterprises	4.90				4.90										
1988	Treasury bonds	Enterprises	3.50					3.47									
	Treasury bonds	Households	5.70					5.73									
	Key construction bonds	Households	3.10				3.05										
	Fiscal bonds	Financial institutions	6.60				6.60										
1989	Treasury bonds	Households	5.60						5.60								
	Price-indexed bonds	Households	12.50						12.50								
	Special state bonds	Enterprises	4.30								4.28						
1990	Treasury bonds	Households	9.30							9.28							
	Special state bonds	Enterprises	3.20									3.20					
	Fiscal bonds	Financial institutions	7.10									7.11					
	Conversion bonds	Enterprises	9.40									9.40					
1991	Treasury bonds	Households	19.94								19.94						
	Special state bonds	Enterprises	2.00										2.00				
	Fiscal bonds	Financial institutions	7.00										7.00				
1992	Treasury bonds	Households	14.90											14.90			
	Treasury bonds	Households	24.70									24.70					
1993	Treasury bonds	Households	22.64										22.64				
	Treasury bonds	Households	8.84												8.84		
	Fiscal bonds	Financial institutions	7.00										7.00				
1994	Treasury bonds	Enterprises	2.00													2.00	
	Treasury bonds	Households	8.24									8.24					
	Treasury bonds	Households	28.50										28.50				
	Treasury bonds	Households	70.00											70.00			
1995	Special state bonds	Financial institutions	3.00														3.00
	Treasury bonds	Financial institutions	25.00												25		
	Treasury bonds	Households	105.88												105.68		
	Treasury bonds	Households	6.00												6.00		
	Treasury bonds	Financial institutions	11.89										11.89				
TOTAL AMOUNT MATURING			474.63	0.87	1.71	2.55	23.53	18.00	26.04	10.12	24.22	52.65	108.67	84.90	145.52	2.00	3.00
	Treasury bonds	Households		0.39	0.81	1.25	5.00	10.55	10.46	9.72	19.94	24.70	22.64	14.90	8.84	0.00	0.00
	Treasury bonds	Enterprises		0.48	0.90	1.30	3.48	6.58	3.08	0.40	0.00	0.00	0.00	0.00	0.00	0.00	0.00
	Key construction bonds	Enterprises		0.00	0.00	0.00	4.90	0.00	0.00	0.00	0.00	0.00	0.00	0.00	0.00	0.00	0.00
	Key construction bonds	Households		0.00	0.00	0.00	3.55	0.00	0.00	0.00	0.00	0.00	0.00	0.00	0.00	0.00	0.00
	Special state bonds	Enterprises		0.00	0.00	0.00	0.00	0.00	0.00	0.00	4.28	3.20	2.00	0.00	0.00	0.00	0.00
	Fiscal bonds	Financial institutions		0.00	0.00	0.00	6.60	0.00	0.00	0.00	0.00	7.11	7.00	0.00	6.00	0.00	0.00
	Conversion bonds	Enterprises		0.00	0.00	0.00	0.00	0.00	0.00	0.00	0.00	9.40	0.00	0.00	0.00	0.00	0.00
	Price-indexed bonds	Households		0.00	0.00	0.00	0.00	0.00	12.50	0.00	0.00	0.00	0.00	0.00	0.00	0.00	0.00

Source : Ministry of Finance.

Table 14: Monetary Survey /a /b

	1993	1994	1995				1996			
			March	June	Sept.	Dec.	March	June	Sept.	Dec.
	(billions of yuan, end of period)									
Net foreign assets	222	506	508	503	541	637	729	801	871	951
Net domestic assets	3,107	3,976	4,290	4,566	4,882	5,188	5,495	5,777	6,084	6,410
Domestic credit	3,301	4,160	4,310	4,410	4,642	5,098	5,265	5,650	5,979	6,195
Loans to enterprises and individuals	3,158	3,937	4,081	4,239	4,449	4,871	5011	5,357	5,510	5,889
Net credit to government	118	131	129	77	100	131	173	212	201	216
Claims on nonmonetary financial institution	25	91	100	94	93	96	81	82	86	89
Other items (net)	-194	-183	-20	156	240	91	230	126	287	215
Money plus quasi-money (broad money)	3,330	4,483	4,797	5,069	5,423	5,825	6,224	6,577	6,955	7,361
Money	1,628	2,054	2,103	2,142	2,248	2,399	2,391	2,462	2,634	2,851
Currency	586	729	727	700	737	789	817	767	841	880
Demand deposits	1,042	1,325	1,376	1,442	1,511	1,610	1,574	1,695	1,793	1,971
Household demand deposits	..	..	..	..	..	..	..	..	..	..
Enterprise deposits	..	..	..	..	..	..	..	..	..	..
Official institutions	..	..	..	..	..	..	..	..	..	..
Quasi-money	1,702	2,429	2,695	2,927	3,175	3,426	3,833	4,115	4,321	4,510
	(twelve-month percentage change)									
Net domestic assets	26.4	28.0	34.6	29.6	31.0	30.5	28.1	26.5	24.6	23.5
of which										
Domestic credit	22.6	26.0	23.5	19.9	19.9	22.5	22.2	28.1	24.9	21.5
Loans to enterprises and individuals	..	..	..	..	..	..	..	..	..	..
Money and quasi-money (broad money)	24.0	34.6	35.4	32.4	30.6	29.9	29.7	29.8	28.2	26.4
of which: currency	35.2	24.3	24.6	21.1	14.9	8.2	12.4	9.8	14.1	11.6

/a Covers the operations of the People's Bank, specialized and universal banks, rural and urban credit cooperatives, and the Agricultural Development Bank.

/b Level data from March 1993 have been revised on the basis of a new statistical methodology that includes an improved accounting system and expanded coverage. Growth rates from 1994 are based on these new statistics.

Source : IMF.

Table 15: Operations of the People's Bank /a
(billions of yuan, end of period)

	1993	1994	1995				1996			
			March	June	Sept.	Dec.	March	June	Sept.	Dec.
Net foreign assets	155	445	499	532	602	667	733	841	873	956
Gold and international financial institutions	12	19	15	13	17	16	15	18	19	23
International reserves converted at exchange rate/b	123	434	487	520	579	611	671	719	792	872
Other	20	-7	-3	-1	5	40	46	105	62	62
Net domestic assets	1,160	1,277	1,260	1,197	1,269	1,409	1418	1,294	1,378	1,733
Domestic credit	1,165	1,230	1,202	1,102	1,160	1,298	1293	1,218	1,214	1,565
Claims on deposit money banks	961	1,045	1,030	999	1,037	1,151	1145	1,094	1,117	1,452
Claims on other financial institutions	25	27	31	20	18	18	17	12	12	12
Claims on central government (net)	111	85	67	8	35	61	62	45	18	36
Claims on nonfinancial sectors	68	73	74	74	70	68	68	67	67	66
Other items (net)	-6	46	58	96	109	111	126	76	164	167
Reserve money	1,315	1,722	1,759	1,730	1,871	2,076	2,151	2,135	2,251	2,689
Annual change, in percent/c	35.5	31.1	29.2	22.3	23.1	20.6	22.3	23.4	20.3	29.5
Liabilities to banks	558	745	789	767	835	932	1013	1,029	1,050	1,384
Required deposits	275	383	413	441	478	511	549	592	633	654
Other deposits	229	303	321	266	298	352	396	369	344	666
Cash in vault	54	60	56	59	59	69	68	68	73	63
Liabilities to nonbanks	766	976	970	963	1,035	1,144	1139	1,106	1,201	1,305
Currency in circulation	586	729	727	700	737	789	817	767	841	880
Deposits of financial institutions other than deposit-money banks	50	61	50	59	74	104	60	65	66	115
Deposits of nonfinancial institutions	129	187	193	204	225	251	262	274	294	310
Memorandum Items:										
Money multiplier /d	2.65	2.73	2.86	3.07	3.04	2.93	3.00	3.19	3.20	2.83
Ratio of excess reserves to deposits (%) /e	9.43	9.15	8.74	7.05	7.23	7.97	8.24	7.23	6.55	10.83

/a Data from March 1993 have been revised on the basis of a new statistical methodology that includes an improved accounting system and expanded coverage.

/b Reserves converted at official rate prior to 1994.

/c In November 1996, a strengthening in the enforcement of reserve requirements over rural credit cooperatives was accomodated by an expansion in reserve money. The growth rate of reserve money excluding this operation was just over 20 percent for the year as a whole.

/d Ratio of broad money, as reported in the banking survey, to reserve money.

/e Ratio of banks' excess reserves to deposits reported in the banking survey.

Source: IMF.

Table 16: Banking Survey, 1993-96/1
(billions of yuan, end of period)

	1993	1994	1995				1996			
			Mar.	June	Sep.	Dec.	Mar.	June	Sep.	Dec.
Net foreign assets	222	506	508	503	542	637	682	753	837	921
Net domestic assets	3,266	4,186	4,522	4,812	5,140	5,438	5,769	6,061	6,367	6,688
Domestic credit	3,481	4,310	4,484	4,604	4,844	5,294	5,612	6,049	6,205	6,641
Claims on Government (net)	118	133	131	79	102	132	175	214	203	218
Claims on nonfinancial sectors	3,363	4,177	4,353	4,525	4,742	5,161	5,437	5,835	6,003	6,423
Other items, net	-215	-125	38	208	295	144	157	12	162	47
Broad money	3,488	4,692	5,030	5,315	5,681	6,075	6,451	6,813	7,204	7,609
Narrow money	1,628	2,054	2,103	2,142	2,249	2,399	2,391	2,462	2,634	2,851
Currency in circulation	586	728	727	700	737	789	817	767	841	880
Demand deposits	1,042	1,325	1,376	1,442	1,512	1,610	1,574	1,695	1,793	1,971
Quasi-money	1,860	2,638	2,927	3,173	3,432	3,676	4,060	4,351	4,571	4,758
Time deposits	125	194	220	265	310	332	372	430	467	504
Savings deposits	1,520	2,152	2,376	2,557	2,757	2,966	3,330	3,546	3,708	3,852
Other deposits	215	292	331	351	365	378	358	375	395	402
					(Twelve-month change, in percent)					
Memorandum items: /2										
Net domestic assets	...	28.2	35.2	30.2	30.9	29.9	27.6	25.9	23.9	23.0
Domestic credit	...	23.8	22.5	19.0	18.8	22.8	25.1	31.4	28.1	25.5
Broad money	...	34.5	35.9	32.7	30.6	29.5	28.3	28.2	26.8	25.3
Narrow money	...	26.2	27.9	21.2	18.3	16.8	13.7	14.9	17.1	18.9
Quasi-money	...	41.8	42.3	41.9	40.1	39.3	38.7	37.1	33.2	29.4
Currency in circulation	35.2	24.3	24.6	21.1	14.9	8.2	12.4	9.4	14.1	11.6

/1 Includes the operations of the People's Bank of China, the deposit money banks, and other banks (or specific depository institutions). Data for March 1996 and later include, in addition, operations of two policy banks (the Export-Import Bank and the State Development Bank).
/2 Owing to a break in the series in 1993, growth rates for that year are not available.

Source : Data provided by the Chinese authorities.

Table 17: Balance Sheets of Urban Credit Cooperatives
(billions of yuan, end of period)

	1993	1994	1995				1996			
			March	June	Sept.	Dec.	March	June	Sept.	Dec.
Foreign assets (net)	0	0	0	0	0	0	0	0	0	0
Reserve assets	33	50	51	53	54	68	62	60	62	78
Required reserves	13	23	25	28	31	33	36	35	37	39
Deposits with the PBC	18	21	20	19	17	28	21	20	19	34
Cash in vault	2	3	4	4	4	5	5	5	6	4
Central bank bonds	0	2	2	2	2	2	0	0	0	0
Claims on central government	7	7	9	11	10	10	12	9	9	10
Claims on other sectors	78	144	162	174	187	207	218	224	241	263
Claims on nonmonetary financial institutions	0	14	15	17	18	19	19	20	21	24
Liabilities to nonfinancial sector	134	235	253	275	295	336	340	341	362	400
Demand deposits	62	100	99	103	105	117	107	99	103	117
Time deposits	7	20	25	30	36	41	41	42	44	47
Savings deposits	38	73	86	97	109	126	144	157	168	183
Other deposits	27	42	42	44	45	52	47	44	47	52
Liabilities to central bank	2	3	4	3	3	3	3	3	3	3
Liabilities to nonmonetary financial institutions	0	4	5	5	5	6	6	8	10	18
Bonds	0	0	0	0	0	0	0	0	0	0
Owners' equity	8	16	18	18	19	20	22	20	20	21
Paid-in capital	8	11	13	13	13	14	16	15	15	16
Other items (net) /a	-27	-45	-42	-45	-52	-61	-61	-58	-62	-67

/a In keeping with the authorities' presentation, "other items, net" is shown as a negative entry on the liabilities side, rather than a positive entry on the assets side and does not net out bonds and owners' equity.

Source : IMF.

Table 18: Balance Sheets of Rural Credit Cooperatives
(billions of yuan, end of period)

	1993	1994	1995				1996			
			March	June	Sept.	Dec.	March	June	Sept.	Dec.
Foreign assets (net)	0	0	0	0	0	0	0	0	0	0
Reserve assets	61	88	91	103	106	122	120	124	129	194
Required reserves	51	67	74	79	83	88	92	98	102	71
Deposits with the PBC	1	9	7	12	11	19	15	12	13	107
Cash in vault	9	12	11	12	12	16	14	14	15	16
Central bank bonds	0	0	0	0	0	0	0	0	0	0
Claims on central government	0	0	0	0	0	0	0	39	40	41
Claims on other sectors	314	417	464	477	502	523	564	595	628	636
Claims on nonmonetary financial institutions	0	0	0	0	0	0	0	0	0	0
Liabilities to nonfinancial sector	430	568	616	649	678	717	773	816	843	879
Demand deposits	68	81	75	77	81	89	79	85	89	101
Time deposits	4	6	6	7	9	9	9	11	11	12
Savings deposits	358	482	535	564	589	620	685	720	743	767
Other deposits	0	0	0	0	0	0	0	0	0	0
Liabilities to central bank	0	0	0	0	0	0	0	0	0	0
Liabilities to nonmonetary financial institutions	0	0	0	0	0	0	0	0	0	0
Bonds	0	0	0	0	0	0	0	0	0	0
Owners' equity	41	62	62	64	64	63	56	55	56	55
Paid-in capital	41	69	37	36	36	38	38	38	38	40
Other items (net) /a	-96	-124	-123	-133	-134	-135	-145	-112	-101	-63

/a In keeping with the authorities' presentation, "other items, net" is shown as a negative entry on the liabilities side, rather than a positive entry on the assets side and does not net out bonds and owners' equity.

Source : IMF.

Table 19: Consolidated Government Revenue /a /b
(billions of yuan)

	1985	1986	1987	1988	1989	1990	1991	1992	1993	1994	1995	1996
Total revenue	228.3	244.6	257.6	280.3	326.4	355.0	367.2	392.8	475.9	558.4	654.1	771.9
Tax revenue	218.8	224.8	232.1	257.6	301.7	313.8	331.7	345.7	447.5	524.0	605.6	691.6
Taxes on income and profits /c	73.9	73.4	71.7	76.4	78.9	84.7	82.1	81.1	80.5	94.0	111.0	129.6
Enterprises income tax	69.7	68.9	66.5	68.2	69.4	74.5	73.1	72.1	67.9	70.9	83.3	92.7
State enterprises	59.6	59.6	56.3	57.1	58.3	60.4	62.7	62.5	58.3	61.0	67.8	77.9
Collectives	10.1	9.3	10.2	10.5	10.5	13.3	10.4	9.6	9.6	9.9	15.5	14.8
Joint ventures	..	..	..	0.6	0.6	0.8	..	..	..	..	..	19.4
Personal income tax (other)	..	..	..	0.8	1.0	1.4	..	..	..	..	..	36.7
Agricultural income tax	4.2	4.5	5.2	7.4	8.5	8.8	9.0	9.0	12.6	23.1	27.7	36.9
Taxes on goods and services	96.3	106.7	111.9	130.8	158.2	164.0	175.5	206.6	287.8	346.5	399.0	461.8
General sales taxes	95.3	104.0	109.5	126.3	144.8	149.7	159.9	205.8	286.8	346.5	399.0	461.8
Product tax	59.4	54.7	53.9	48.1	53.0	58.1	62.9	69.3	82.1	48.7	54.0	61.6
Value added tax	14.8	23.2	25.4	38.4	43.1	40.0	40.6	70.6	108.1	230.8	258.6	295.9
Business tax	21.1	26.1	30.2	39.8	48.7	51.6	56.4	65.9	96.6	67.0	86.4	104.3
Urban maintenance and development tax	..	..	..	..	8.6	9.2	10.0	..	..	..	..	..
Real estate tax	..	..	..	2.2	2.6	3.2	3.8	..	..	..	..	..
Special tax on oil	..	1.6	1.5	1.4	1.2	1.1	1.0	..	0.2	..	..	..
Salt tax	1.0	1.1	0.9	0.9	1.0	0.8	0.8	0.8	0.8	..	..	..
Customs tax	20.5	15.2	14.2	15.5	18.2	15.9	18.7	21.3	25.6	27.3	28.7	30.2
Other taxes	28.1	29.5	34.3	34.9	46.4	49.2	55.4	36.7	53.6	56.2	66.9	70.0
of which construction tax	2.3	2.4	3.0	2.6	2.8	3.8	3.1	3.2	3.8	4.3	5.3	6.3
Nontax revenue	9.5	19.8	25.5	22.7	24.7	41.2	35.5	47.1	28.4	34.4	48.5	80.3
Gross profit remittances from state owned enterprises /d	4.4	4.2	4.3	5.1	6.4	7.8	7.5	6.0	4.9	..	..	..
Depreciation funds	..	..	..	..	..	..	..	..	..	..	..	..
Other	5.1	15.6	21.2	17.6	18.3	33.4	28.0	41.1	23.5	34.4	48.5	80.3
of which foreign grants (net)	0.2	0.4	-0.1	0.0	0.5	0.2	2.1	1.9	1.7	4.3	5.0	2.5
Memorandum item:												
Gross profit remittances from state owned enterprises	4.4	4.2	4.3	5.1	6.4	7.8	7.5	6.0	4.9	..	..	..
GNP in current prices	898.9	1,020.1	1,195.5	1,492.2	1,691.8	1,859.8	2,166.3	2,665.2	3,456.1	4,653.3	5,727.7	6,693.4

/a According to the definition contained in IMF, *Manual on Government Finance Statistics* (GFS), 1986.
/b This includes all government revenue, with the exception of extrabudgetary receipts of the various levels of government.
/c Beginning with 1985, profit taxes on state enterprises are included under tax revenue.
/d As of 1988, only banks and financial institutions are subject to remittance.
Source : IMF.

Table 20: Structure of Consolidated Government Revenue
(percentage of total revenue)

	1985	1986	1987	1988	1989	1990	1991	1992	1993	1994	1995	1996
Total revenue	100.0	100.0	100.0	100.0	100.0	100.0	100.0	100.0	100.0	100.0	100.0	100.0
Tax revenue	95.8	91.9	90.1	91.9	92.4	88.4	90.3	88.0	94.0	93.8	92.6	89.6
Taxes on income and profits	32.4	30.0	27.8	27.3	24.2	23.9	22.4	20.6	16.9	16.8	17.0	16.8
Enterprises income tax	30.5	28.2	25.8	24.3	21.3	21.0	19.9	18.4	14.3	12.7	12.7	12.0
State enterprises	26.1	24.4	21.9	20.4	17.9	17.0	17.1	15.9	12.3	10.9	10.4	10.1
Collectives	4.4	3.8	4.0	3.7	3.2	3.7	2.8	2.4	2.0	1.8	2.4	1.9
Joint ventures	..	..	..	0.2	0.2	0.2	..	..	..	..	..	2.5
Personal income tax (other)	..	..	..	0.3	0.3	0.4	..	..	..	..	..	4.8
Agricultural income tax	1.8	1.8	2.0	2.6	2.6	2.5	2.5	2.3	2.6	4.1	4.2	4.8
Taxes on goods and services	42.2	43.6	43.4	46.7	48.5	46.2	47.8	52.6	60.5	62.1	61.0	59.8
General sales taxes	41.7	42.5	42.5	45.1	44.4	42.2	43.5	52.4	60.3	62.1	61.0	59.8
Product tax	26.0	22.4	20.9	17.2	16.2	16.4	17.1	17.6	17.3	8.7	8.3	8.0
Value added tax	6.5	9.5	9.9	13.7	13.2	11.3	11.1	18.0	22.7	41.3	39.5	38.3
Business tax	9.2	10.7	11.7	14.2	14.9	14.5	15.4	16.8	20.3	12.0	13.2	13.5
Urban maintenance and development tax	..	..	..	..	2.6	2.6	2.7	..	..	..	..	..
Real estate tax	..	..	..	0.8	0.8	0.9	1.0	..	..	..	..	..
Special tax on oil	..	0.7	0.6	0.5	0.4	0.3	0.3	..	..	..	..	..
Salt tax	0.4	0.4	0.3	0.3	0.3	0.2	0.2	0.2	0.2	..	..	..
Customs tax	9.0	6.2	5.5	5.5	5.6	4.5	5.1	5.4	5.4	4.9	4.4	3.9
Other taxes	12.3	12.1	13.3	12.5	14.2	13.9	15.1	9.3	11.3	10.1	10.2	9.1
of which:												
Construction tax	1.0	1.0	1.2	0.9	0.9	1.1	0.8	0.8	0.8	0.8	0.8	0.8
Nontax revenue	4.2	8.1	9.9	8.1	7.6	11.6	9.7	12.0	6.0	6.2	7.4	10.4
Gross profit remittances from state enterprises	1.9	1.7	1.7	1.8	2.0	2.2	2.0	1.5	1.0	0.0	0.0	0.0
Depreciation funds	..	..	..	..	..	..	..	..	..	..	..	..
Other	2.2	6.4	8.2	6.3	5.6	9.4	7.6	10.5	4.9	6.2	7.4	10.4
of which foreign grants (net)	0.1	0.2	0.0	0.0	0.2	0.1	0.6	0.5	0.3	0.8	0.8	0.3
Memorandum item:												
Gross profit remittances from state owned enterprises	1.9	1.7	1.7	1.8	2.0	2.2	2.0	1.5	1.0	..	..	..

Source: Table 19.

Table 21: Structure of Government Revenue (percentage of GNP)

	1985	1986	1987	1988	1989	1990	1991	1992	1993	1994	1995	1996
Total revenue	25.4	24.0	21.5	18.8	19.3	19.1	17.0	14.7	13.8	12.0	11.4	11.5
Tax revenue	24.3	22.0	19.4	17.3	17.8	16.9	15.3	13.0	12.9	11.3	10.6	10.3
Taxes on income and profits	8.2	7.2	6.0	5.1	4.7	4.6	3.8	3.0	2.3	2.0	1.9	1.9
Enterprises income tax	7.8	6.8	5.6	4.6	4.1	4.0	3.4	2.7	2.0	1.5	1.5	1.4
State enterprises	6.6	5.8	4.7	3.8	3.4	3.2	2.9	2.3	1.7	1.3	1.2	1.2
Collectives	1.1	0.9	0.9	0.7	0.6	0.7	0.5	0.4	0.3	0.2	0.3	0.2
Joint ventures	..	..	..	..	..	..	..	..	..	..	..	0.3
Personal income tax (other)	..	..	..	0.1	0.1	0.1	..	..	..	..	..	0.5
Agricultural income tax	0.5	0.4	0.4	0.5	0.5	0.5	0.4	0.3	0.4	0.5	0.5	0.6
Taxes on goods and services	10.7	10.5	9.4	8.8	9.4	8.8	8.1	7.8	8.3	7.4	7.0	6.9
General sales taxes	10.6	10.2	9.2	8.5	8.6	8.0	7.4	7.7	8.3	7.4	7.0	6.9
Product tax	6.6	5.4	4.5	3.2	3.1	3.1	2.9	2.6	2.4	1.0	0.9	0.9
Value added tax	1.6	2.3	2.1	2.6	2.5	2.2	1.9	2.6	3.1	5.0	4.5	4.4
Business tax	2.3	2.6	2.5	2.7	2.9	2.8	2.6	2.5	2.8	1.4	1.5	1.6
Urban maintenance and development tax	..	..	..	..	0.5	0.5	0.5	..	..	..	..	..
Real estate tax	..	..	..	0.1	0.2	0.2	0.2	..	..	..	..	..
Special tax on oil	..	0.2	0.1	0.1	0.1	0.1	..	..	..	..	..	..
Salt tax	0.1	0.1	0.1	0.1	0.1	..	..	..	..	..	..	..
Customs tax	2.3	1.5	1.2	1.0	1.1	0.9	0.9	0.8	0.7	0.6	0.5	0.5
Other taxes	3.1	2.9	2.9	2.3	2.7	2.6	2.6	1.4	1.6	1.2	1.2	1.0
of which:												
Construction tax	0.3	0.2	0.3	0.2	0.2	0.2	0.1	0.1	0.1	0.1	0.1	0.1
Nontax revenue	1.1	1.9	2.1	1.5	1.5	2.2	1.6	1.8	0.8	0.7	0.8	1.2
Gross profit remittances from state owned enterprises	0.5	0.4	0.4	0.3	0.4	0.4	0.3	0.2	0.1	0.0	0.0	0.0
Depreciation funds	..	..	..	..	..	..	..	..	..	..	..	..
Other	0.6	1.5	1.8	1.2	1.1	1.8	1.3	1.5	0.7	0.7	0.8	1.2
of which foreign grants (net)	0.0	0.0	0.0	0.0	0.0	0.0	0.1	0.1	0.0	0.1	0.1	..
Memorandum item:												
Gross profit remittances from state owned enterprises	0.5	0.4	0.4	0.3	0.4	0.4	0.3	0.2	0.1	..	..	..

Source: Table 19.

Table 22: Structure of Government Expenditure

	1985	1986	1987	1988	1989	1990	1991	1992	1993	1994	1995	1996
	(billions of yuan)											
Total expenditure and net lending	232.4	262.7	282.7	313.7	363.8	391.7	415.2	453.9	545.9	632.5	751.8	875.7
Current expenditure	168.1	187.5	207.1	237.6	288.2	306.9	328.5	364.0	424.3	538.4	624.0	734.1
Administration	14.4	18.2	19.5	23.9	28.5	33.3	34.3	42.5	53.6	72.9	87.2	105.5
Defense	19.2	20.1	21.0	21.8	25.1	29.0	33.0	37.8	42.6	55.1	63.7	72.0
Culture, education, public health, science & broadcasting	31.7	38.0	40.3	48.6	55.3	61.7	70.8	79.3	95.8	127.8	146.6	170.5
of which: education	18.4	21.4	22.7	27.9	31.6	35.3	41.0	45.3	55.8	..	..	..
Economic services	22.4	25.2	26.0	28.9	34.6	37.9	46.0	47.8	57.4	57.8	55.0	63.0
Geological survey	3.0	3.1	3.0	3.3	3.3	3.6	7.8	4.4	4.9	6.4	6.6	6.9
Agriculture	10.1	12.4	13.4	15.4	19.7	22.2	24.4	26.9	32.3	40.0	44.9	51.8
Operating expenditure for industry, communication & commerce	3.5	3.7	3.3	3.9	4.5	4.7	5.2	6.5	7.6			
Development of new products	4.4	5.0	5.1	5.3	5.9	6.3	7.3	8.9	10.7	11.4	0.0	0.0
Working capital for state owned enterprises	1.4	1.0	1.2	1.0	1.2	1.1	1.3	1.1	1.9	..	3.5	4.3
Social welfare relief	..	3.6	3.7	4.1	5.0	5.5	6.7	6.6	7.5	9.5	11.5	15.2
Subsidies	50.7	58.2	67.0	76.3	97.3	96.0	88.3	76.7	71.0	68.0	71.7	79.4
Daily living necessities	31.4	25.7	29.5	31.7	37.4	38.1	37.3	32.2	29.9	31.4	36.4	44.1
Agricultural inputs	1.3	..	..	..	..	..	..	..	..	..	..	..
Operating losses of state owned enterprises	18.0	32.5	37.5	44.6	59.9	57.9	51.0	44.5	41.1	36.6	35.3	35.3
Interest payments	..	1.9	2.8	3.0	2.8	4.4	7.8	14.2	9.7	16.7	35.6	48.9
Other	29.7	22.3	26.7	31.0	39.6	39.1	41.6	59.1	86.7	130.6	152.7	179.6
Capital expenditure	64.3	75.2	75.6	76.1	75.6	84.8	86.7	89.9	121.6	94.1	127.8	141.6
Capital construction	58.4	67.2	68.2	66.4	66.9	75.8	76.0	76.5	90.1	64.0	79.4	88.6
Development of the productive capacity of existing enterprises	5.9	8.0	7.4	9.7	8.7	9.0	10.7	13.4	31.5	30.1	48.4	53.0
	(percentage of GNP)											
Memorandum items:												
Current expenditure	18.7	18.4	17.3	15.9	17.0	16.5	15.2	13.7	12.3	13.4	12.8	11.0
Subsidies	5.6	5.7	5.6	5.1	5.8	5.2	4.1	2.9	2.1	1.5	1.3	1.2
Daily living necessities	3.5	2.5	2.5	2.1	2.2	2.0	1.7	1.2	0.9	0.7	0.6	0.7
Operating losses of state owned enterprises	2.0	3.2	3.1	3.0	3.5	3.1	2.4	1.7	1.2	0.8	0.6	0.5
Capital expenditure	7.2	7.4	6.3	5.1	4.5	4.6	4.0	3.4	3.5	2.0	2.2	2.1
	(as a percentage of total expenditure)											
Subsidies	21.8	22.2	23.7	24.3	26.7	24.5	21.3	16.9	13.0	10.8	9.5	9.1
Capital expenditure	27.7	28.6	26.7	24.3	20.8	21.6	20.9	19.8	22.3	14.9	17.0	16.2

Source: IMF.

Table 23: Budget and Its Financing

	1985	1986	1987	1988	1989	1990	1991	1992	1993	1994	1995	1996
(billions of yuan)												
Revenue	228.3	244.6	257.6	280.3	326.4	355.0	367.2	392.8	475.9	558.4	654.1	771.9
Expenditure	232.4	262.7	282.7	313.7	363.8	391.7	415.2	453.9	545.9	632.4	751.8	875.6
Deficit	-4.1	-18.1	-25.1	-33.4	-37.4	-36.7	-48.0	-61.1	-70.0	-74.2	-97.7	-103.7
Financing	4.1	18.1	25.1	33.4	37.3	36.8	48.0	61.2	70.0	74.2	97.7	103.7
Domestic	10.1	12.8	18.3	22.0	26.0	23.7	38.0	46.1	41.9	67.0	101.0	103.7
Banking system	4.0	12.8	7.9	22.0	26.0	23.7	16.1	88.9	-31.6	14.9	-0.9	85.7
Nonbank	6.1	0.0	10.4	0.0	0.0	0.0	21.9	-42.8	73.5	52.2	101.9	18.0
Foreign	-6.0	5.3	6.8	11.4	11.3	13.1	10.0	15.1	28.1	7.1	-3.3	0.0
Gross foreign borrowing	..	7.6	10.6	13.9	14.4	17.8	18.0	20.9	35.8	14.7	1.1	11.6
Amortization	..	-2.3	-3.8	-2.5	-3.1	-4.7	-8.0	-5.8	-7.7	-7.6	-4.4	-11.6
(percentage of GNP)												
Revenue	25.4	24.0	21.5	18.8	19.3	19.1	17.0	14.7	13.8	12.0	11.4	11.5
Expenditure	25.9	25.7	23.6	21.0	21.5	21.1	19.2	17.0	15.8	13.6	13.1	13.1
Deficit	-0.5	-1.8	-2.1	-2.2	-2.2	-2.0	-2.2	-2.3	-2.0	-1.6	-1.7	-1.6
Financing	0.5	1.8	2.1	2.2	2.2	2.0	2.2	2.3	2.0	1.6	1.7	1.6
Domestic	1.1	1.3	1.5	1.5	1.5	1.3	1.8	1.7	1.2	1.4	1.8	1.6
Banking system	0.4	1.3	0.7	1.5	1.5	1.3	0.7	3.3	-0.9	0.3	0.0	1.3
Nonbank	0.7	0.0	0.9	0.0	0.0	0.0	1.0	-1.6	2.1	1.1	1.8	0.3
Foreign	-0.7	0.5	0.6	0.8	0.7	0.7	0.5	0.6	0.8	0.2	-0.1	0.0
Gross foreign borrowing	..	0.7	0.9	0.9	0.9	1.0	0.8	0.8	1.0	0.3	0.0	0.2
Amortization	..	-0.2	-0.3	-0.2	-0.2	-0.3	-0.4	-0.2	-0.2	-0.2	-0.1	-0.2
(percentage of total deficit)												
Financing	100.0	100.2	100.4	99.7	99.8	100.3	100.0	100.2	100.0	100.0	100.0	100.0
Domestic	246.4	70.9	73.1	65.8	69.5	64.6	79.2	75.5	59.9	90.5	103.4	100.0
Banking system	97.6	70.9	31.5	65.8	69.5	64.6	33.5	145.5	-45.2	20.1	-0.9	82.6
Nonbank	148.8	0.0	41.5	0.0	0.0	0.0	45.6	-70.0	105.0	70.4	104.3	17.4
Foreign	-146.4	29.3	27.3	34.0	30.3	35.7	20.8	24.7	40.2	9.6	-3.4	0.0
Gross foreign borrowing	..	42.1	42.5	41.4	38.5	48.5	37.5	34.2	51.2	19.8	1.1	11.2
Amortization	..	-12.7	-15.2	-7.5	-8.2	-12.8	-16.7	-9.5	-11.0	-10.3	-4.5	-11.2

Source: IMF.

Table 24: Production of Major Crops

	1985	1986	1987	1988	1989	1990	1991	1992	1993	1994	1995
						(millions of tons)					
Total food grains	379.11	391.51	402.98	394.08	407.45	446.24	435.29	442.66	456.49	445.10	466.62
Rice	168.57	172.22	174.26	169.11	180.13	189.33	183.81	186.22	177.70	175.93	185.23
Wheat	85.81	90.04	85.90	85.43	90.81	98.23	95.95	101.59	106.39	99.30	102.21
Corn	63.88	70.86	79.24	77.35	78.93	96.82	98.77	95.38	102.70	99.28	111.99
Tuber	26.04	25.34	28.20	26.97	27.30	27.43	27.16	28.44	31.81	30.25	32.63
Total oil seeds	15.78	14.74	15.28	13.20	12.91	16.13	16.38	16.41	18.04	19.90	22.50
of which:											
Peanuts	6.66	5.88	6.17	5.69	5.36	6.37	6.30	5.95	8.42	9.68	10.24
Rapeseed	5.61	5.88	6.61	5.04	5.44	6.96	7.44	7.65	6.94	7.49	9.78
Cotton	4.15	3.54	4.25	4.15	3.79	4.51	5.68	4.51	3.74	4.34	4.77
Sugarcane	51.55	50.22	47.36	49.06	48.57	57.62	67.90	73.01	64.19	60.93	65.42
Beetroots	8.92	8.31	8.14	12.81	9.36	14.53	16.29	15.07	12.05	12.53	13.98
Cured tobacco	2.08	1.37	1.64	2.34	2.41	2.26	2.67	3.12	3.01	1.94	2.07
						(millions of tons)					
Fruits	11.64	13.48	16.68	16.66	18.37	18.74	21.76	24.40	30.11	35.00	42.15
Apples	3.61	3.34	4.26	4.34	4.50	4.32	4.54	6.56	9.07	11.13	14.01
Citrus	1.81	2.55	3.22	2.56	4.56	4.86	6.33	5.16	6.56	6.81	8.22
Pears	2.14	2.35	2.49	2.72	2.57	2.35	2.50	2.85	3.22	4.04	4.94
Bananas	0.63	1.25	2.03	1.83	1.40	1.46	1.98	2.45	2.70	2.90	3.13

Source: *China Statistical Yearbook* 1996, pp. 371.

Table 25: Yield of Major Crops
(by sown area, kg/hectare)

	1985	1986	1987	1988	1989	1990	1991	1992	1993	1994	1995
Total food grains	3,480	3,525	3,615	3,585	3,630	3,930	3,870	4,004	..	..	..
Rice	5,250	5,340	5,415	5,280	5,505	5,730	5,640	5,803	..	..	..
Wheat	2,940	3,045	2,985	2,970	3,045	3,195	3,105	3,331	..	..	..
Corn	3,600	3,705	3,915	3,930	3,885	4,530	4,575	4,533	..	..	..
Soybeans	1,365	1,395	1,470	1,440	1,275	1,455	1,380	1,427	..	..	..
Tuber	3,030	2,910	3,180	2,985	3,000	3,015	2,985	3,141	..	..	..
Peanuts	2,010	1,815	2,040	1,905	1,815	2,190	2,190	2,000	2,492	2,564	2,687
Rapeseed	1,245	1,200	1,260	1,020	1,095	1,260	1,215	1,281	1,309	1,296	1,416
Cotton	810	825	870	750	735	810	870	660	750	785	879
Sugarcane	53,430	52,860	55,140	53,115	50,850	57,120	58,350	58,605	59,012	57,671	58,136
Beetroots	15,915	15,960	16,350	17,190	16,245	21,660	20,790	22,832	20,124	17,936	20,132
Cured tobacco	1,920	1,530	1,785	1,800	1,605	1,680	1,710	1,687	1,654	1,491	1,584

Source: China Statistical Yearbook 1996 , pp. 374.

Table 26: Gross Output Value of Industry

	1985	1986	1987	1988	1989	1990	1991	1992	1993	1994	1995
					(billions of yuan)						
Total	971.6	1,119.4	1,381.3	1,822.5	2,201.7	2,392.4	2,662.5	3,459.9	4,840.2	7,017.6	9,189.5
BY TYPE OF OWNERSHIP											
State-owned	630.2	697.1	825.0	1,035.1	1,234.3	1,306.4	1,495.5	1,782.4	2,272.5	2,620.1	3,122.0
Collective-owned	311.7	375.2	478.2	658.7	785.8	852.3	878.3	1,213.5	1,646.4	2,647.2	3,362.3
Township	76.1	98.1	128.4	184.7	219.4	244.1	240.1	353.4	537.4	810.2	1,193.2
Village	66.3	83.8	116.5	170.4	211.8	239.4	234.7	363.2	516.3	965.8	1,184.7
Joint urban-rural	15.2	24.8	31.6	43.9	49.6	53.9	56.9	87.0	132.2	261.1	213.4
Joint urban	0.0	2.1	3.0	3.9	5.0	5.5	6.9	10.2	15.6	33.8	..
Joint rural	15.2	22.7	28.6	40.0	44.6	48.4	50.0	76.8	116.6	227.3	..
Individual-owned	18.0	30.9	50.2	79.0	105.8	129.0	128.7	200.6	386.1	708.2	1,182.1
Urban	3.3	2.9	5.0	6.8	9.0	10.7	12.9	19.5	39.6	86.7	..
Rural	14.6	27.9	45.2	72.2	96.8	118.3	..	..	..	..	..
Other	11.7	16.3	27.9	49.5	75.8	104.8	163.1	268.8	517.4	901.8	1,523.1
BY TYPE OF INDUSTRY											
Light	457.5	533.0	665.6	897.9	1,076.1	1,181.3	1,380.1	1,749.2	2,318.4	2,167.1	2,349.1
Heavy	514.1	586.4	715.7	924.5	1,125.6	1,211.3	1,444.7	1,957.4	2,950.8	2,968.2	3,145.6
					(percentage of total)						
BY TYPE OF OWNERSHIP											
State-owned	64.9	62.3	59.7	56.8	56.1	54.6	56.2	51.5	46.9	37.3	34.0
Collective-owned	32.1	33.5	34.6	36.1	35.7	35.6	33.0	35.1	34.0	37.7	36.6
Township	7.8	8.8	9.3	10.1	10.0	10.2	9.0	10.2	11.1	11.5	13.0
Village	6.8	7.5	8.4	9.3	9.6	10.0	8.8	10.5	10.7	13.8	12.9
Joint urban-rural	1.6	2.2	2.3	2.4	2.3	2.3	2.1	2.5	2.7	3.7	2.3
Joint urban	0.0	0.2	0.2	0.2	0.2	0.2	0.3	0.3	0.3	0.5	..
Joint rural	1.6	2.0	2.1	2.2	2.0	2.0	1.9	2.2	2.4	3.2	..
Individual-owned	1.8	2.8	3.6	4.3	4.8	5.4	4.8	5.8	8.0	10.1	12.9
Urban	0.3	0.3	0.4	0.4	0.4	0.4	0.5	0.6	0.8	1.2	..
Rural	1.5	2.5	3.3	4.0	4.4	4.9	..	..	..	..	..
Other	1.2	1.5	2.0	2.7	3.4	4.4	6.1	7.8	10.7	12.9	16.6
BY TYPE OF INDUSTRY											
Light	47.1	47.6	48.2	49.3	48.9	49.4	51.8	50.6	47.9	30.9	25.6
Heavy	52.9	52.4	51.8	50.7	51.1	50.6	54.3	56.6	61.0	42.3	34.2

Note : For 1991-1994 different categories do not add up to total.
Source: China Statistical Yearbook 1996, pp. 401.

Table 27: Output of Major Industrial Products

Product	1985	1986	1987	1988	1989	1990	1991	1992	1993	1994	1995
Coal (million tons)	872	894	928	980	1,054	1,080	1,087	1,116	1,150	1,240	1,361
Crude oil (million tons)	125	131	134	137	137	138	141	142	145	146	150
Natural gas (billion cu m)	13	14	14	14	15	15	16	16	17	18	18
Electricity (billion kWh)	411	450	497	545	585	621	678	754	840	928	1,008
Hydro power	92	95	100	109	118	127	125	131	152	167	191
Steel (million tons)	47	52	56	59	62	66	71	81	90	93	95
Rolled steel (million tons)	37	41	44	47	49	52	56	67	77	84	90
Cement (million tons)	146	166	186	210	210	210	253	308	368	421	476
Timber (million cu m)	63	65	64	62	58	56	58	62	64	66	60
Fertilizers (million tons)	13	14	17	17	18	19	20	20	20	23	26
(growth rates)											
Coal	10.5	2.5	3.8	5.6	7.6	2.5	0.6	2.7	3.0	7.8	9.8
Crude oil	9.0	4.6	2.6	2.0	0.1	1.0	1.9	0.8	2.2	0.6	2.7
Natural gas	4.0	6.4	0.9	2.7	5.5	1.7	5.1	-1.8	6.2	4.7	2.2
Electricity	8.9	9.4	10.6	9.6	7.3	6.2	9.1	11.3	11.4	10.6	8.6
Hydro power	6.5	2.5	5.6	9.2	8.3	7.1	-1.6	4.8	16.1	10.3	13.8
Steel	7.6	11.6	7.8	5.6	3.6	7.7	7.0	14.0	10.6	3.4	3.0
Rolled steel	9.5	9.9	8.1	7.1	3.6	5.9	9.4	18.8	15.2	9.2	6.5
Cement	18.6	13.8	12.2	12.8	0.1	-0.3	20.5	22.0	19.4	14.5	13.0
Timber	-1.0	2.8	-1.4	-3.0	-6.6	-4.0	4.2	6.3	3.5	3.5	-8.6
Fertilizers	-9.5	2.8	23.0	4.1	4.6	3.3	5.3	3.5	-4.5	16.2	12.5

Source: China Statistical Yearbook 1996 , pp. 433.

Table 28: Total Wage Bill of Staff and Workers by Employment Category

	1985	1986	1987	1988	1989	1990	1991	1992	1993	1994	1995
					(billions of yuan)						
Total wage bill	138.3	166.0	188.1	231.6	261.9	295.1	332.4	393.9	491.6	665.6	810.0
State-owned	106.5	128.9	145.9	180.7	205.0	232.4	259.5	309.0	381.3	517.7	608.0
Urban collectives	31.2	36.3	40.9	48.8	53.4	58.1	65.9	74.3	85.0	102.3	118.2
Other	0.6	0.8	1.3	2.2	3.4	4.6	7.0	10.6	25.4	45.6	63.8
Total wage bill	100.0	100.0	100.0	100.0	100.0	100.0	100.0	100.0	100.0	100.0	97.5
State-owned	77.0	77.6	77.6	78.0	78.3	78.8	78.1	78.5	77.6	77.8	75.1
Urban collectives	22.6	21.9	21.7	21.1	20.4	19.7	19.8	18.9	17.3	15.4	14.6
Other	0.4	0.5	0.7	0.9	1.3	1.6	2.1	2.7	5.2	6.8	7.9
					(growth rates)						
Total wage bill	22.0	20.0	13.3	23.1	13.1	12.7	12.6	18.5	24.8	35.4	21.7
State-owned	21.6	21.0	13.3	23.8	13.5	13.4	11.7	19.1	23.4	35.8	17.4
Urban collectives	23.0	16.2	12.8	19.2	9.6	8.7	13.4	12.8	14.4	20.4	15.5
Other	63.9	42.4	50.0	70.6	57.7	35.7	53.0	50.0	140.2	79.7	40.0
Percentage share of wage bill in GDP	15.4	16.3	15.7	15.5	15.5	15.9	15.4	14.8	14.2	14.3	13.9

Source: China Statistical Yearbook 1996, pp. 115.

Table 29: Average Annual Wage by Sector and Employment Category (yuan in current prices)

Sector	1985	1986	1987	1988	1989	1990	1991	1992	1993	1994	1995
Staff and workers											
Total	1,148	1,329	1,459	1,747	1,935	2,140	2,340	2,711	3,371	4,538	5,500
Farming & forestry, etc.	878	1,048	1,143	1,280	1,389	1,541	1,652	1,828	2,042	2,819	3,522
Excavation	1,324	1,569	1,663	1,964	2,378	2,718	2,942	3,209	3,711	4,679	5,757
Manufacturing	1,112	1,275	1,418	1,710	1,900	2,073	2,289	2,635	3,348	4,283	5,169
Electric power, gas,and water	1,239	1,497	1,677	1,971	2,241	2,656	2,922	3,392	4,319	6,155	7,843
Geological survey	1,406	1,604	1,768	2,025	2,199	2,465	2,707	3,222	3,717	5,450	5,962
Construction	1,362	1,536	1,684	1,959	2,166	2,384	2,649	3,066	3,779	4,894	5,785
Transport & communications	1,275	1,476	1,621	1,941	2,197	2,426	2,686	3,114	4,273	5,690	6,948
Commerce & services etc.	1,007	1,148	1,270	1,556	1,660	1,818	1,981	2,204	2,679	3,537	4,248
Real estate	1,028	1,216	1,327	1,715	1,925	2,243	2,507	3,106	4,320	6,288	7,330
Social services	777	980	1,085	1,719	1,926	2,170	2,431	2,844	3,588	5,026	5,982
Health care, sports & welfare	1,124	1,343	1,446	1,752	1,959	2,209	2,370	2,812	3,413	5,126	5,860
Education, culture & arts etc.	1,166	1,330	1,409	1,747	1,883	2,117	2,243	2,715	3,278	4,923	5,435
Scientific research	1,272	1,492	1,620	1,931	2,118	2,403	2,573	3,115	3,904	6,162	6,846
Banking and insurance	1,154	1,353	1,458	1,739	1,867	2,097	2,255	2,829	3,740	6,712	7,376
Government agencies	1,127	1,356	1,468	1,707	1,874	2,113	2,275	2,768	3,505	4,962	5,526
Others									3,371	5,213	6,295
Staff and workers in state-owned enterprises											
Total	1,213	1,414	1,546	1,853	2,055	2,284	2,477	2,878	3,532	4,797	5,625
Farming & forestry, etc.	892	1,062	1,154	1,291	1,401	1,559	1,665	1,845	2,043	2,821	3,527
Excavation	1,384	1,638	1,734	2,038	2,449	2,763	2,982	3,239	3,856	4,863	5,944
Manufacturing	1,190	1,382	1,543	1,872	2,081	2,289	2,505	2,889	3,562	4,508	5,352
Electric power, gas,and water	1,272	1,518	1,692	1,994	2,248	2,648	2,883	3,354	4,317	6,124	7,734
Geological survey	1,408	1,607	1,773	2,025	2,199	2,463	2,718	3,235	3,729	5,476	5,987
Construction	1,532	1,731	1,882	2,192	2,419	2,667	2,924	3,406	4,182	5,498	6,512
Transport & communications	1,383	1,610	1,773	2,140	2,423	2,697	2,967	3,452	4,604	6,212	7,572
Commerce & services etc.	1,087	1,268	1,398	1,737	1,851	2,028	2,201	2,478	2,933	3,856	4,568
Real estate	1,170	1,364	1,500	1,750	1,992	2,247	2,476	3,082	4,278	5,997	6,884
Social services	1,208	1,417	1,545	1,842	2,028	2,307	2,547	3,008	3,661	5,098	5,949
Health care, sports & welfare	1,164	1,376	1,481	1,793	1,999	2,263	2,417	2,883	3,494	5,267	6,009
Education, culture & arts etc.	1,184	1,344	1,422	1,764	1,899	2,134	2,257	2,732	3,292	4,944	5,457
Scientific research	1,268	1,494	1,624	1,935	2,123	2,411	2,580	3,130	3,898	6,212	6,835
Banking and insurance	1,234	1,427	1,540	1,842	1,960	2,200	2,355	2,967	3,885	7,017	7,595
Government agencies	1,133	1,361	1,472	1,709	1,875	2,115	2,277	2,774	3,512	4,967	5,528
Others									3,793	5,744	6,854
Staff and workers in urban collective-owned enterprises											
Total	967	1,092	1,207	1,426	1,557	1,681	1,866	2,109	2,592	3,245	3,931
Farming & forestry, etc.	725	875	979	1,111	1,178	1,238	1,366	1,487	1,887	2,510	2,927
Excavation	852	945	1,016	1,208	1,433	1,844	1,960	2,000	2,327	2,793	3,680
Manufacturing	963	1,075	1,180	1,388	1,523	1,622	1,798	2,017	2,469	3,076	3,717
Electric power, gas,and water	667	875	1,250	1,294	1,625	2,133	2,588	2,737	3,539	5,734	7,461
Geological survey	900	1,286	1,333	1,500	1,000	1,212	1,765	2,188	2,843	3,692	4,294
Construction	1,101	1,232	1,380	1,597	1,763	1,935	2,216	2,554	3,182	3,936	4,677
Transport & communications	1,009	1,139	1,236	1,409	1,575	1,661	1,854	2,070	2,711	3,110	3,584
Commerce & services etc.	912	1,008	1,118	1,335	1,417	1,548	1,691	1,827	2,213	2,823	3,449
Real estate	1,050	1,238	1,130	1,602	1,967	1,969	2,432	2,763	4,006	5,290	6,706
Social services	806	1,057	1,167	1,333	1,521	1,638	1,905	2,082	2,727	3,754	4,707
Health care, sports & welfare	975	1,212	1,296	1,570	1,774	1,956	2,135	2,416	2,935	4,238	4,890
Education, culture & arts etc.	779	945	1,033	1,202	1,352	1,533	1,689	1,987	2,539	3,548	4,291
Scientific research	1,052	1,286	1,319	1,636	1,710	1,997	2,120	2,392	3,474	4,719	6,046
Banking and insurance	945	1,150	1,235	1,450	1,597	1,806	1,965	2,428	3,182	5,625	6,407
Government agencies	1,046	1,227	1,368	1,648	1,860	2,042	2,206	2,565	3,071	4,411	5,314
Others									2,547	4,067	4,935

Source: China Statistical Yearbook 1996, pp.117-124.

Table 30: Social Labor Force by Sector (millions of workers)

	1985	1986	1987	1988	1989	1990	1991	1992	1993	1994	1995
Farming forestry, animal husbandry, fishery & water c	311.3	312.5	316.6	322.5	332.3	341.2	349.6	348.0	339.7	333.9	330.2
Excavation	8.0	8.1	8.2	8.3	8.4	8.8	9.1	9.0	9.3	9.2	9.3
Manufacturing	74.1	80.2	83.6	86.5	85.5	86.2	88.4	91.1	93.0	96.1	98.0
Electric power, gas and water	1.4	1.5	1.6	1.8	1.8	1.9	2.0	2.2	2.4	2.5	2.6
Geological survey & exploration	2.0	2.0	2.0	2.0	2.0	2.0	2.0	2.0	1.4	1.4	1.4
Construction	20.4	22.4	23.8	24.9	24.1	24.2	24.8	26.6	30.5	31.9	33.2
Transportation, posts & telecommunications	12.8	13.8	14.5	15.2	15.2	15.7	16.2	16.7	16.9	18.6	19.4
Commerce, catering trade, supply & marketing of materials and warehouses	23.1	24.1	25.8	27.4	27.7	28.4	30.0	32.1	34.6	39.2	42.9
Real estate	0.4	0.4	0.4	0.4	0.4	0.4	0.5	0.5	0.7	0.7	0.8
Social services	4.0	4.7	5.0	5.3	5.5	5.9	6.0	6.4	5.4	6.3	7.0
Public health, sports and social welfare	4.7	4.8	5.0	5.1	5.2	5.4	5.5	5.7	4.2	4.3	4.4
Education, culture, art, radio and television broadcasting	12.7	13.2	13.8	14.0	14.3	14.6	15.0	15.2	12.1	14.4	14.8
Scientific research, technical service	1.4	1.5	1.6	1.6	1.7	1.7	1.8	1.8	1.7	1.8	1.8
Banking and insurance	1.4	1.5	1.7	1.9	2.1	2.2	2.3	2.5	2.7	2.6	2.8
Governments, parties and organizations	8.0	8.7	9.3	9.7	10.2	10.8	11.4	11.5	10.3	10.3	10.4
Others	13.2	13.4	15.0	16.6	17.1	18.0	19.1	23.1	37.4	41.6	44.9
TOTAL	498.7	512.8	527.8	543.4	553.3	567.4	583.6	594.3	602.2	614.7	623.9
Agriculture	311.3	312.5	316.6	322.5	332.3	341.2	349.6	348.0	339.7	333.9	330.2
Industry	103.8	112.2	117.3	121.5	119.8	121.2	124.3	128.8	135.2	139.6	143.2
Service	83.6	88.1	93.9	99.4	101.3	105.0	109.8	117.6	127.4	141.2	150.6

Note: Agriculture consists of farming, forestry, animal husbandry, fishery and water conservancy; services are residual.

Source: *China Statistical Yearbook* 1996, pp. 92.

Table 31: Social Labor Force by Employment Category

	1985	1986	1987	1988	1989	1990	1991	1992	1993	1994	1995
					(millions of workers)						
Total	498.7	512.8	527.8	543.4	553.3	567.4	583.6	594.3	602.2	614.7	623.9
							583.6				
Staff and workers	123.6	128.1	132.1	136.1	137.4	140.6	145.1	147.9	148.5	152.6	153.0
State-owned	89.9	93.3	96.5	99.8	101.1	103.5	106.6	108.9	109.2	112.1	112.6
Urban collective-owned	33.2	34.2	34.9	35.3	35.0	35.5	36.3	36.2	33.9	32.9	31.5
Other ownership	0.4	0.6	0.7	1.0	1.3	1.6	2.2	2.8	5.4	7.6	8.9
Urban individual laborers	4.5	4.8	5.7	6.6	6.5	6.7	7.6	8.4	11.2	15.6	20.5
Rural laborers	370.7	379.9	390.0	400.7	409.4	420.1	430.9	438.0	442.6	446.5	450.4
					(percentage of total)						
Staff and workers	24.8	25.0	25.0	25.0	24.8	24.8	24.9	24.9	24.7	24.8	24.5
State-owned	18.0	18.2	18.3	18.4	18.3	18.2	18.3	18.3	18.1	18.2	18.0
Urban collective-owned	6.7	6.7	6.6	6.5	6.3	6.3	6.2	6.1	5.6	5.3	5.0
Other ownership	0.1	0.1	0.1	0.2	0.2	0.3	0.4	0.5	0.9	1.2	1.4
Urban individual laborers	0.9	0.9	1.1	1.2	1.2	1.2	1.3	1.4	1.9	2.5	3.3
Rural laborers	74.3	74.1	73.9	73.7	74.0	74.0	73.8	73.7	73.5	72.6	72.2
					(growth rates)						
Total	3.5	2.8	2.9	2.9	1.8	2.5	2.9	1.8	1.3	2.1	1.5
Staff and workers	3.9	3.6	3.2	3.0	1.0	2.3	3.2	2.0	0.4	2.8	0.3
State-owned	4.1	3.8	3.4	3.4	1.3	2.3	3.1	2.1	0.3	2.7	0.4
Urban collective-owned	3.4	2.9	2.0	1.1	-0.7	1.3	2.2	-0.2	-6.3	-3.2	-4.2
Other ownership	18.9	25.0	30.9	34.7	36.1	24.2	31.7	30.6	91.5	40.7	17.6
Urban individual laborers	32.7	7.3	17.8	15.8	-1.7	3.4	13.4	10.5	33.3	39.3	31.4
Rural laborers	3.0	2.5	2.7	2.7	2.2	2.6	2.6	1.7	1.0	0.9	0.9

Source: China Statistical Yearbook 1996, p. 90.

Table 32: General Price Indices
(annual growth rates)

	1985	1986	1987	1988	1989	1990	1991	1992	1993	1994	1995	1996
Overall Retail Price	8.8	6.0	7.3	18.5	17.8	2.1	2.9	5.4	13.2	21.7	14.8	6.2
Overall Retail Price Index 1990=100	61.7	65.4	70.2	83.1	97.9	100.0	102.9	108.5	122.8	149.4	171.5	182.1
Overall Consumer Price of which	9.3	6.5	7.3	18.8	18.0	3.1	3.4	6.4	14.7	24.1	17.1	8.3
Urban	11.9	7.0	8.8	20.7	16.3	1.3	5.1	8.6	16.1	25.0	16.8	8.8
Rural	7.6	6.1	6.2	17.5	19.3	4.5	2.3	4.7	13.7	23.4	17.5	7.9
Overall Farm and Sideline Purchasing Price	8.6	6.4	12.0	23.0	15.0	-2.6	-2.0	3.4	13.4	39.9	19.9	..
Overall Industrial Products Rural Retail Price	3.2	3.2	4.8	15.2	18.7	4.6	3.0	3.1	11.8	17.2	14.7	..
Overall Industrial and Agricultural Products Price Parity	-5.0	-3.0	-6.4	-6.3	3.2	7.4	5.1	-0.3	-1.4	-16.2	-4.4	..

Source: China Statistical Yearbook 1996, p. 255; China Monthly Statistics.

Table 33: Annual Rates of Inflation by Month
(percentage change in the index over the previous 12 months)

	JAN	FEB	MAR	APR	MAY	JUN	JUL	AUG	SEP	OCT	NOV	DEC
Overall Retail Price index												
1987	5.0	5.1	5.5	6.5	7.6	7.8	8.0	8.4	7.9	7.6	8.5	9.1
1988	9.5	11.2	11.6	12.6	14.7	16.5	19.3	23.2	25.4	26.1	26.0	26.7
1989	27.0	27.9	26.3	25.8	24.3	21.5	19.0	15.2	11.4	8.7	7.1	6.4
1990	4.1	4.1	3.3	3.1	2.6	3.0	0.7	0.4	0.8	1.1	1.6	2.2
1991	1.4	1.0	0.9	0.6	3.1	3.8	4.2	4.0	4.3	4.0	4.4	4.0
1992	5.1	4.9	5.0	6.2	4.1	4.2	4.3	4.7	5.7	6.4	6.6	6.7
1993	8.4	8.7	10.2	10.9	12.5	13.9	14.9	15.1	14.5	14.6	15.1	17.6
1994	18.8	20.9	20.2	19.5	18.9	19.6	21.4	23.5	24.6	25.2	25.0	23.3
1995	21.2	19.7	18.7	18.0	17.6	16.0	14.6	12.3	11.4	10.3	9.2	8.3
1996	7.6	7.7	7.7	7.4	6.5	5.9	5.8	5.8	5.0	4.7	4.6	6.1
1997	3.3	2.9										
Cost of Living Index												
1987	5.0	5.3	5.6	6.6	7.7	7.9	8.0	8.4	7.8	7.6	8.5	9.2
1988	9.7	10.7	11.6	12.6	14.5	16.7	19.8	24.1	26.4	27.0	26.8	27.7
1989	27.4	28.4	27.0	26.5	24.7	22.5	18.9	14.8	10.8	7.7	6.4	5.5
1990	3.5	3.5	2.7	2.5	2.0	2.4	0.2	-0.1	0.5	0.9	1.4	2.1
1991	1.5	1.0	0.8	0.4	3.0	3.8	4.3	4.1	4.4	4.0	4.6	4.0
1992	5.1	4.9	5.2	6.9	4.3	4.3	4.7	5.1	5.9	6.7	6.8	7.0
1993	8.6	8.9	10.2	10.4	11.9	13.3	14.3	14.6	14.1	14.4	15.2	17.9
1994	21.1	23.2	22.4	21.7	21.3	22.6	24.0	25.8	27.4	27.7	27.5	25.5
1995	24.1	22.4	21.3	20.7	20.3	18.2	16.7	14.5	13.2	12.1	11.2	10.1
1996	9.0	9.3	9.8	9.7	8.9	8.6	8.3	8.1	7.4	7.0	6.9	7.0
1997	5.9	5.6										
Free market Price index of Consumer Goods												
1987	22.0	7.9	7.4	14.2	15.6	20.1	16.9	20.0	16.5	19.9	22.6	20.8
1988	20.0	27.9	28.8	23.1	21.5	22.4	27.7	33.2	37.3	36.5	34.8	30.6
1989	27.5	26.9	19.2	21.0	20.1	15.0	14.9	9.1	3.3	0.9	-6.4	-6.6
1990	-5.1	-6.6	-5.1	-2.0	-5.7	-5.3	-7.3	-7.5	-4.4	-4.3	-3.8	-0.5
1991	-2.3	-0.8	-5.2	-6.9	-5.5	0.3	1.8	0.6	-0.2	1.7	1.0	0.8
1992	3.7	-0.3	2.0	1.6	2.5	-1.5	-0.2	2.5	3.1	6.5	3.7	3.1
1993	6.7	4.1	4.6	8.0	10.0	14.4	16.0	14.8	14.4	11.8	16.7	23.6

Source: China Monthly Statistics .

Table 34: Total Investment in Fixed Assets

	1985	1986	1987	1988	1989	1990	1991	1992	1993	1994	1995
						(billions of yuan)					
Total fixed ii	254.3	302.0	364.1	449.7	413.8	444.9	550.9	785.5	1,245.8	1,704.3	2,001.9
State-ow	168.1	197.9	229.8	276.3	253.5	273.3	337.8	476.8	765.8	961.6	1,089.8
Collectiv	32.7	39.2	54.7	71.2	57.0	52.9	69.8	135.9	223.1	275.9	328.9
Individuɛ	53.5	64.9	79.6	102.2	103.2	100.1	118.3	122.2	147.6	197.1	256.0
Other	0.0	0.0	0.0	0.0	0.0	18.6	25.0	50.6	109.2	269.8	327.1
Fixed investment by CENTER & LOCAL											
CENTEF	..	..	..	105.8	116.8	128.0	150.5	184.7	..	..	..
LOCAL	..	..	..	170.5	136.7	145.3	187.3	277.5	..	..	..
						(percentage of total)					
Total fixed ii	100.0	100.0	100.0	100.0	100.0	100.0	100.0	100.0	100.0	100.0	100.0
State-ow	66.1	65.5	63.1	61.4	61.3	61.4	61.3	60.7	61.5	56.4	54.4
Collectiv	12.9	13.0	15.0	15.8	13.8	11.9	12.7	17.3	17.9	16.2	16.4
Individuɛ	21.0	21.5	21.9	22.7	24.9	22.5	21.5	15.6	11.8	11.6	12.8
Other	0.0	0.0	0.0	0.0	0.0	4.2	4.5	6.4	8.8	15.8	16.3
Fixed investment by CENTER & LOCAL											
CENTEF	..	..	..	38.3	46.1	46.8	44.5	38.7	..	..	..
LOCAL	..	..	..	61.7	53.9	53.2	55.5	58.2	..	..	..
Total fixed ii	28.4	29.6	30.4	30.1	24.5	24.0	25.5	29.5	36.0	36.6	34.4

Source: China Statistical Yearbook 1996, p.140 for 1995; previous issues for earlier years.

Table 35: Investment in Fixed Assets by State Owned Enterprises

	1985	1986	1987	1988	1989	1990	1991	1992	1993	1994	1995
					(billions of yuan)						
Fixed Investment	168.1	197.9	229.8	276.3	253.5	273.3	337.8	476.8	765.7	961.5	1089.8
Capital Construction	107.4	117.6	134.3	157.4	155.2	170.4	211.6	301.3	461.6	643.7	740.4
Technical Upgdating	44.9	61.9	75.9	98.1	78.9	83.0	102.3	146.1	219.6	291.9	329.9
Other	15.7	18.3	19.6	20.8	19.5	19.9	23.9	29.4	84.6	26.0	19.5
Fixed Investment by Sector of which	168.1	197.9	229.8	276.3	253.5	273.3	337.8	476.8	765.7	961.5	1089.8
Raw materials	..	..	..	69.5	64.0	71.8	85.9	68.8	..	..	..
Energy	..	..	..	64.5	70.6	82.4	95.7	114.9	..	..	..
Transport and communications	..	..	..	28.5	22.9	26.9	40.7	75.7	..	..	..
Construction incl. geography	..	..	..	2.8	2.2	1.8	2.1	3.9	..	..	..
Real Estate	..	..	..	18.8	14.8	11.3	16.7	26.5	..	..	..
					(percentage of total)						
Fixed Investment	100.0	100.0	100.0	100.0	100.0	100.0	100.0	100.0	100.0	100.0	100.0
Capital Construction	63.9	59.4	58.4	57.0	61.2	62.3	62.6	63.2	60.3	66.9	67.9
Technical Upgdating	26.7	31.3	33.0	35.5	31.1	30.4	30.3	30.6	28.7	30.4	30.3
Other	9.3	9.3	8.5	7.5	7.7	7.3	7.1	6.2	11.0	2.7	1.8
Fixed Investment by Sector	100.0	100.0	100.0	100.0	100.0	100.0	100.0	100.0	100.0	100.0	100.0
of which											
Raw materials	..	..	..	25.2	25.2	26.3	25.4	14.4	..	..	..
Energy	..	..	..	23.3	27.8	30.1	28.3	24.1	...	...	...
Transport and communications	..	..	..	10.3	9.0	9.8	12.1	15.9	..	..	..
Construction incl. geography	..	..	..	1.0	0.8	0.7	0.6	0.8	..	..	..
Real Estate	..	..	..	6.8	5.9	4.1	4.9	5.6	..	..	..

Source: China Statistical Yearbook 1996, p.148 for 1985-95.

Table 36: Investment in Capital Constuction of State-Owned Enterprises by Sector

	1985	1986	1987	1988	1989	1990	1991	1992	1993	1994	1995
						(billions of yuan)					
All Sectors	107.4	117.6	134.3	157.4	155.2	170.4	211.6	301.3	461.6	643.7	740.4
Agriculture	1.7	1.6	2.0	2.3	2.0	2.6	3.3	4.4	4.6	5.7	7.7
Industry	44.6	53.2	68.3	81.3	82.2	95.3	114.7	145.8	200.4	276.2	323.6
Geology	2.6	2.6	3.0	2.9	3.6	4.6	5.9	8.2	9.8	12.1	16.6
Construction	2.2	1.9	1.5	1.5	1.4	1.0	1.3	2.3	11.5	13.8	14.6
Transport	17.8	18.8	19.5	21.6	17.0	21.1	34.0	45.8	90.1	137.3	158.8
Commerce	4.0	3.5	4.2	5.0	4.2	3.9	6.4	13.7	20.3	25.5	24.9
Real Estate & Social Services	11.8	11.0	9.3	13.4	11.2	8.2	12.2	21.0	44.3	73.4	67.3
Educaiton, Health, & Culture etc.	10.1	11.6	12.6	13.2	12.9	13.8	15.2	19.7	27.1	35.5	45.8
Research	2.1	2.5	2.6	2.3	2.2	2.1	2.3	3.2	4.9	5.3	6.8
Banking & Insurance	0.7	0.8	1.3	1.9	1.6	1.5	1.9	3.0	6.7	9.6	12.6
Government Agencies & Other	9.8	10.1	10.0	12.0	17.0	16.3	14.4	34.3	41.8	49.4	61.8
						(percentage of total)					
All Sectors	100.0	100.0	100.0	100.0	100.0	100.0	100.0	100.0	100.0	100.0	100.0
Agriculture	1.6	1.3	1.5	1.5	1.3	1.5	1.6	1.4	1.0	0.9	1.0
Industry	41.6	45.2	50.8	51.6	53.0	55.9	54.2	48.4	43.4	42.9	43.7
Geology	2.5	2.2	2.3	1.8	2.3	2.7	2.8	2.7	2.1	1.9	2.2
Construction	2.0	1.6	1.1	1.0	0.9	0.6	0.6	0.8	2.5	2.1	2.0
Transport	16.6	16.0	14.5	13.7	11.0	12.4	16.1	15.2	19.5	21.3	21.4
Commerce	3.7	2.9	3.1	3.2	2.7	2.3	3.0	4.5	4.4	4.0	3.4
Real Estate & Social Services	11.0	9.4	6.9	8.5	7.2	4.8	5.8	7.0	9.6	11.4	9.1
Educaiton, Health, & Culture etc.	9.4	9.9	9.3	8.4	8.3	8.1	7.2	6.5	5.9	5.5	6.2
Research	1.9	2.2	2.0	1.5	1.4	1.2	1.1	1.0	1.1	0.8	0.9
Banking & Insurance	0.7	0.7	1.0	1.2	1.0	0.9	0.9	1.0	1.4	1.5	1.7
Government Agencies & Other	9.1	8.6	7.5	7.6	10.9	9.6	6.8	11.4	9.1	7.7	8.3

Source: China Statistical Yearbook 1996, p.148.

Table 37: Foreign Direct Investment Inflows
(millions of US dollars)

	1985	1986	1987	1988	1989	1990	1991	1992	1993	1994	1995
TOTAL	1,958.7	2,243.7	2,646.6	3,739.7	3,773.5	3,754.9	4,666.6	11,291.6	27,770.9	33,945.8	37,805.7
of which, from											
Hong Kong & Macao	955.7	1,328.7	1,809.1	2,428.1	2,341.8	2,118.5	2,661.8	7,908.9	18,032.5	20,332.1	20,624.9
Japan	315.1	263.4	266.6	598.4	407.7	520.5	609.5	748.3	1,361.4	2,086.2	3,212.5
Korea	..	..	..	..	..	..	..	674.8	381.5	726.1	1,047.1
Taiwan (China)	..	..	..	..	..	..	471.9	1,053.4	3,139.1	3,391.3	3,165.2
United Kingdom	71.4	35.3	13.8	46.6	29.0	19.9	37.9	38.5	220.5	688.8	915.2
France	32.5	43.6	17.3	31.6	11.6	23.4	11.7	46.9	141.4	193.4	287.0
Italy	19.4	29.4	21.5	36.2	34.2	8.1	41.3	26.7	99.9	206.2	270.2
United States	357.2	326.2	271.3	244.4	288.2	461.2	330.7	519.4	2,067.9	2,490.8	3,083.7
RECEIVED BY PROVINCES											
Regional Total	1,320.6	1,741.7	1,782.7	3,149.7	3,437.3	3,436.4	4,425.8	11,003.3	27,341.7	33,267.7	37,215.5
Beijing	88.8	149.7	105.8	503.2	320.2	279.0	245.0	349.9	666.9	1,371.6	1,080.0
Tianjin	63.8	134.8	133.1	61.2	31.4	36.9	132.6	107.8	613.7	1,015.0	1,520.9
Hebei	55.9	51.4	10.3	19.1	43.7	44.5	56.6	113.1	396.5	523.4	546.7
Shanxi	0.5	0.2	4.9	6.5	9.8	3.4	3.8	53.8	86.4	31.7	63.8
Inner Mongolia	2.6	7.5	5.1	6.4	4.4	10.6	1.7	5.2	85.3	40.1	57.8
Liaoning	25.8	48.2	90.8	130.6	126.1	257.3	362.4	516.4	1,279.1	1,440.1	1,424.6
Jilin	4.9	24.2	7.4	9.7	99.9	17.6	31.6	75.3	275.3	241.9	408.0
Heilongjiang	4.0	24.5	14.0	69.3	57.4	28.4	20.9	72.2	232.3	347.6	516.9
Shanghai	108.8	148.9	214.0	233.2	422.1	174.0	145.2	493.6	3,160.3	2,473.1	2,892.6
Jiangsu	51.1	45.6	86.4	125.5	126.9	134.0	219.2	1,463.2	2,843.7	3,763.2	5,190.8
Zhejiang	26.6	24.8	36.3	43.8	54.0	49.1	92.3	239.8	1,031.8	1,150.3	1,258.1
Anhui	3.0	35.2	3.2	27.9	8.8	13.5	10.7	54.7	257.6	370.0	482.6
Fujian	118.6	62.5	55.4	145.5	348.0	319.9	471.2	1,423.6	2,874.4	3,713.2	4,043.9
Jiangxi	10.5	9.1	5.4	8.9	9.2	7.5	19.5	99.7	208.2	261.7	288.9
Shandong	35.6	65.7	65.0	89.7	163.3	185.7	216.4	1,003.4	1,874.1	2,552.4	2,689.0
Henan	8.3	10.7	13.5	64.2	46.1	11.4	38.0	53.2	304.9	386.7	478.6
Hubei	8.0	12.4	26.0	22.3	28.6	31.8	46.6	203.1	540.5	601.9	625.1
Hunan	31.0	28.5	2.9	12.9	23.3	14.2	25.4	132.7	437.5	331.1	507.7
Guangdong	651.3	862.7	736.9	1,251.1	1,323.2	1,582.3	1,942.9	3,701.1	7,555.8	9,463.4	10,260.1
Guangxi	30.9	49.2	45.1	20.9	53.0	35.6	31.9	182.0	884.6	836.3	672.6
Hainan	..	..	..	117.4	95.0	103.0	176.7	452.6	707.1	918.1	1,062.1
Sichuan	28.7	31.8	24.3	40.3	13.1	24.4	80.9	112.1	571.4	921.7	541.6
Guizhou	9.8	12.2 ..		13.8	13.9	11.1	16.3	19.8	42.9	63.6	57.0
Yunnan	1.6	3.8	6.3	8.3	7.9	7.4	3.5	28.8	97.0	65.0	97.7
Tibet	..	..	..	..	..	..	..	..	..	..	..
Shaanxi	15.6	37.2	72.9	111.7	97.2	47.3	31.8	45.5	234.3	238.8	324.1
Gansu	0.6	1.3	0.2	2.4	0.0	1.2	4.8	0.4	12.0	87.8	63.9
Qinghai	0.2	0.0	0.0	2.7	0.0	0.0	0.0	0.7	3.2	2.4	1.6
Ningxia	0.3	0.1	0.0	0.3	1.1	0.3	0.2	0.4	11.9	7.3	3.9
Xinjiang	10.9	14.0	17.7	5.0	0.9	5.4	0.2	0.0	53.0	48.3	54.9

Source: China Statistical Yearbook 1996, pp. 598-600 for 1994-95; previous issues for earlier years.

Table 38: Production and Consumption of Energy

	1985	1986	1987	1988	1989	1990	1991	1992	1993	1994	1995
PRODUCTION (millions of tons of coal equivalent)	855.5	881.2	912.7	958.0	1,016.4	1,039.2	1,048.4	1,072.6	1,110.6	1,187.3	1,287.3
					(percentage of total)						
Coal	72.8	72.4	72.6	73.1	74.1	74.2	74.1	74.3	74.0	74.6	75.5
Crude oil	20.9	21.2	21.0	20.4	19.3	19.0	19.2	18.9	18.7	17.6	16.7
Natural gas	2.0	2.1	2.0	2.0	2.0	2.0	2.0	2.0	2.0	1.9	1.8
Hydro power	4.3	4.3	4.4	4.5	4.6	4.8	4.7	4.8	5.3	5.9	6.0
CONSUMPTION	766.8	808.5	866.3	930.0	969.3	987.0	1,037.8	1,091.7	1,159.9	1,227.4	1,290.0
					(percentage of total)						
Coal	75.8	75.8	76.2	76.2	76.0	76.2	76.1	75.7	74.7	75.0	75.0
Crude oil	17.1	17.2	17.0	17.0	17.2	16.6	17.1	17.5	18.2	17.4	17.3
Natural gas	2.2	2.3	2.1	2.1	2.0	2.1	2.0	1.9	1.9	1.9	1.8
Hydro power	4.9	4.7	4.7	4.7	4.9	5.1	4.8	4.9	5.2	5.7	5.9
GDP (billion of yuan, constant 1990 price)	1,136.7	1,255.7	1,421.8	1,595.9	1,654.3	1,721.7	1,874.3	2,149.9	2,401.0	2,726.6	3,016.3
Energy consumption (million ton per billion yuan)	0.7	0.6	0.6	0.6	0.6	0.6	0.6	0.5	0.5	0.5	0.4

Notes: Excluding bio-energy, solar, geothermal and nuclear energy. All fuels are converted into standard fuel with thermal equivalent of 7,000 kilocalories per kilogram. The conversion is
1 kg of coal (5,000 kcal) = 0.714 kg of standard fuel.
1 kg of crude oil (10,000 kcal) = 1.43 kg of standard fuel.
1 cubic meter of natural gas (9,310 kcal) = 1.33 kg. of standard fuel.
The conversion of hydropower into standard fuel is calculated on the basis of the consumption quota of standard coal for thermal power generation for the year.

Source: China Statistical Yearbook 1996, p. 203.

Table 39: Freight Traffic

	1985	1986	1987	1988	1989	1990	1991	1992	1993	1994	1995
						(billion ton/km)					
Rail	812.6	876.5	947.1	987.8	1,039.4	1,062.2	1,097.2	1,157.6	1,195.5	1,245.8	1,287.0
Road	190.3	211.8	266.0	322.0	337.5	335.8	342.8	375.5	407.1	448.6	469.5
Domestic waterways	240.0	270.0	288.9	310.4	349.8	345.1	396.5	422.2	472.7	1,568.7	1,755.2
Pipelines	60.3	61.2	62.5	65.0	62.9	62.7	62.1	61.7	60.8	61.2	59.0
Civil aviation	0.4	0.5	0.7	0.7	0.7	0.8	1.0	1.3	1.7	1.9	2.2
OVERALL	1,303.7	1,420.0	1,565.2	1,686.0	1,790.3	1,806.6	1,899.7	2,018.4	2,137.7	3,326.1	3,573.0
Ocean shipping	532.9	594.8	657.6	696.6	768.9	814.1	899.0	903.4	913.4	..	..
						(percentage of total)					
Rail	62.3	61.7	60.5	58.6	58.1	58.8	57.8	57.4	55.9	37.5	36.0
Road	14.6	14.9	17.0	19.1	18.9	18.6	18.0	18.6	19.0	13.5	13.1
Domestic waterways	18.4	19.0	18.5	18.4	19.5	19.1	20.9	20.9	22.1	47.2	49.1
Pipelines	4.6	4.3	4.0	3.9	3.5	3.5	3.3	3.1	2.8	1.8	1.7
Civil aviation	0.0	0.0	0.0	0.0	0.0	0.0	0.1	0.1	0.1	0.1	0.1
						(growth rates)					
Rail	12.1	7.9	8.1	4.3	5.2	2.2	3.3	5.5	3.3	4.2	3.3
Road	23.9	11.3	25.6	21.0	4.8	-0.5	2.1	9.6	8.4	10.2	4.6
Domestic waterways	22.4	12.5	7.0	7.5	12.7	-1.3	14.9	6.5	12.0	231.9	11.9
Pipelines	5.4	1.5	2.1	4.0	-3.2	-0.3	-1.0	-0.6	-1.5	0.7	-3.6
Civil aviation	33.4	15.9	35.1	12.3	-5.5	18.8	23.2	32.9	23.8	11.9	20.0
OVERALL	15.2	8.9	10.2	7.7	6.2	0.9	5.1	6.3	5.9	55.6	7.4
Ocean shipping	21.8	11.6	10.6	5.9	10.4	5.9	10.4	0.5	1.1	..	..

Source: China Statistical Yearbook 1996, p. 504.

Table 40: Passenger Traffic

	1985	1986	1987	1988	1989	1990	1991	1992	1993	1994	1995
						(billion passenger/km)					
Rail	241.6	258.7	284.3	326.0	303.7	261.3	282.8	315.2	348.3	363.6	354.6
Road	172.5	198.2	219.0	252.8	266.2	262.0	287.2	319.3	370.1	422.0	460.3
Domestic waterways	17.9	18.2	19.6	20.4	18.8	16.5	17.7	19.8	19.6	18.4	17.2
Civil aviation	11.7	14.6	18.2	21.7	18.7	23.0	30.1	40.6	47.8	55.2	68.1
OVERALL	443.6	489.7	541.1	620.9	607.5	562.8	617.8	694.9	785.8	859.1	900.2
						(as a percentage of total)					
Rail	54.5	52.8	52.5	52.5	50.0	46.4	45.8	45.4	44.3	42.3	39.4
Road	38.9	40.5	40.5	40.7	43.8	46.6	46.5	45.9	47.1	49.1	51.1
Domestic waterways	4.0	3.7	3.6	3.3	3.1	2.9	2.9	2.9	2.5	2.1	1.9
Civil aviation	2.6	3.0	3.4	3.5	3.1	4.1	4.9	5.8	6.1	6.4	7.6
OVERALL	100.0	100.0	100.0	100.0	100.0	100.0	100.0	100.0	100.0	100.0	100.0
						(growth rates)					
Rail	18.1	7.1	9.9	14.7	-6.8	-14.0	8.2	11.5	10.5	4.4	-2.5
Road	29.0	14.9	10.5	15.4	5.3	-1.6	9.6	11.2	15.9	14.0	9.1
Domestic waterways	16.4	1.9	7.6	4.1	-7.7	-12.4	7.5	11.9	-1.0	-6.6	-6.4
Civil aviation	39.8	25.4	24.4	19.2	-13.9	23.4	30.7	34.8	17.6	15.5	23.5
OVERALL	22.5	10.4	10.5	14.7	-2.2	-7.3	9.8	12.5	13.1	9.3	4.8

Source: China Statistical Yearbook 1996, p. 503.

Table 41: Average Shipping Distance

	1985	1986	1987	1988	1989	1990	1991	1992	1993	1994	1995
						(kilometers)					
Rail	636	646	673	681	686	705	718	734	735	791	807
Road	31	34	37	44	46	46	46	48	48	50	50
Waterways	1,216	1,042	1,174	1,128	1,281	1,447	1,554	1,433	1,415	1,465	1,551
Pipelines	442	413	413	416	402	398	399	417	409	406	386
Civil aviation	2,128	2,143	2,183	2,226	2,226	2,218	2,234	2,330	2,393	2,241	2,206
OVERALL	243	236	234	243	258	270	284	279	274	282	289
						(growth rates)					
Rail	8.9	1.6	4.2	1.2	0.7	2.8	1.8	2.2	0.1	7.6	2.0
Road	40.9	9.7	8.8	18.9	4.5	0.0	0.0	4.3	0.0	4.2	0.0
Waterways	-10.0	-14.3	12.7	-3.9	13.6	13.0	7.4	-7.8	-1.3	3.5	5.9
Pipelines	-3.1	-6.6	0.0	0.7	-3.4	-1.0	0.3	4.5	-1.9	-0.7	-4.9
Civil aviation	2.6	0.7	1.9	2.0	0.0	-0.4	0.7	4.3	2.7	-6.4	-1.6
OVERALL	11.0	-2.9	-0.8	3.8	6.2	4.7	5.2	-1.8	-1.8	2.9	2.5

Source: China Statistical Yearbook 1996, p. 505.